The Hidden World Beneath: An Exploration of Soil Fungi

Authored by

Dr. Ravindra Goswami

Dr. Vishwajeet Singh

Dr. Anita Chauhan

Dr. Rashmi Vamil

Prof. Seema Bhadauria

To the unseen architects of our world, the humble fungi, whose quiet work sustains life above and below the surface.

And to all the researchers, ecologists, and curious minds who continue to explore the wonders beneath our feet.

May your discoveries inspire future generations to protect and cherish the hidden world beneath us.

Table of contents

Acknowledgments

Writing this book has been a journey of discovery, and it would not have been possible without the support and encouragement of many individuals and organizations.

First and foremost, I would like to express my deepest gratitude to the scientific community whose tireless research on soil fungi has inspired much of the content in these pages. Your work continues to illuminate the critical role fungi play in our ecosystems, and I am privileged to share some of your insights here.

I am particularly grateful to my co-authors Dr. Vishwajeet Singh, Dr. Anita Chauhan and Dr. Rashmi Vamil who contributed invaluable guidance and expertise throughout the writing process. And special thanks to Prof. Seema Bhadauria, my mentor, whose expertise and feedback have sharpened my understanding and strengthened this book.

Special thanks go to the institutions and laboratories that supported my research, providing access to essential resources, data, and facilities. To the librarians and archivists who assisted with the retrieval of key references—thank you for your patience and dedication.

I am also indebted to my friends and family for their unwavering support. Your encouragement kept me grounded, even when the writing process seemed overwhelming. To those who patiently listened to my ramblings about soil fungi, thank you for your understanding and enthusiasm.

Finally, this book would not have come to fruition without the support of my publisher and the dedicated team behind this project. Your belief in the importance of this work has been instrumental in bringing it to life.

To all who have been part of this journey, my sincerest thanks.

Dr. Ravindra Goswami

Preface

The world beneath our feet is often overlooked, hidden from sight and, for many, from thought. Yet, the soil is teeming with life—a vibrant ecosystem that plays a crucial role in the health of our planet. Among the many organisms that call the soil home, fungi are some of the most remarkable and yet least understood. These silent workers are the architects of soil health, nutrient cycling, plant support, and even carbon sequestration. Despite their critical role in maintaining life on Earth, soil fungi remain shrouded in mystery for most people. This book, *The Hidden World Beneath: An Exploration of Soil Fungi*, seeks to illuminate the fascinating world of fungi that live below ground and their profound impact on our ecosystems.

The idea for this book was born out of both curiosity and necessity. As a researcher and nature enthusiast, I have always been intrigued by the connections between organisms in ecosystems. However, I found that while much attention is given to plants and animals, the fungi—particularly those inhabiting the soil—are often an afterthought. Yet, soil fungi are the foundation upon which healthy ecosystems are built. This book is an attempt to bridge that gap, offering a comprehensive look at the diversity, function, and importance of these incredible organisms.

A Journey into the Soil

Writing this book has been as much an exploration for me as I hope it will be for you. From examining the earliest studies of fungi to diving into the latest breakthroughs in fungal genomics, I have come to appreciate the complexity and versatility of soil fungi. They are not just decomposers breaking down organic matter; they are also symbionts forming partnerships with plants, pathogens challenging agricultural productivity, and unsung heroes in the fight against climate change. Their multifaceted roles in both natural ecosystems and human-influenced environments cannot be overstated.

Throughout this book, I and my co-authors aim to present a balanced view of soil fungi by covering various aspects of their biology, ecology, and applications. Whether you're a student of biology, a farmer interested in sustainable practices, or a curious reader fascinated by the natural world, my goal is to provide insights that deepen your understanding of fungi and their place in the web of life.

Why Fungi Matter Now More Than Ever

In today's world, where climate change, soil degradation, and biodiversity loss are pressing global issues, the role of soil fungi becomes ever more important. Fungi are key players in carbon sequestration and nutrient cycling, processes that are vital for mitigating the effects of global warming. They also form symbiotic relationships with most terrestrial plants, enhancing their ability to absorb water and nutrients, which is crucial in maintaining the health of forests, grasslands, and agricultural systems.

Yet, despite their importance, fungal diversity is under threat. Habitat destruction, pollution, and unsustainable agricultural practices are leading to a decline in fungal populations, which could have far-reaching consequences for ecosystem health and human wellbeing. By understanding the intricacies of soil fungi, we can better appreciate the need for their conservation and the opportunities they offer for sustainable agriculture, environmental remediation, and even medicine.

Structure of the Book

This book is divided into nine chapters, each focusing on different aspects of soil fungi. We begin with an introduction to the basic biology and classification of fungi before moving into the specific types of fungi that inhabit the soil. From there, we explore the critical roles fungi play in nutrient cycling and their symbiotic relationships with plants. Later chapters delve into the impact of soil fungi on agriculture and the environment, as well as their response to global challenges like climate change.

One of the unique aspects of this book is its emphasis on both the scientific and practical implications of soil fungi. While we cover the latest research and findings, real-world examples, and potential applications of fungal knowledge in agriculture and conservation. This combination is intended to make the book accessible to both academic readers and those seeking practical insights.

Acknowledging the Path Ahead

The study of soil fungi is still in its infancy compared to other biological fields, but it is growing rapidly. Advancements in molecular biology, genetics, and ecological modeling are opening new frontiers in our understanding of fungi and their complex roles in ecosystems. However, there is still much we do not know. As you read through this book, I hope it sparks your curiosity to learn more and even contribute to this exciting field of study.

A Personal Note of Gratitude

I would like to take a moment to thank all the researchers and scientists who have dedicated their careers to studying fungi and the soil ecosystems they inhabit. Without your work, this book would not exist. I am also deeply grateful to my co-authors, family, friends, and colleagues who supported me through the writing process. Your encouragement has been invaluable.

Lastly, to the reader: Thank you for picking up this book. I hope that through these pages, you will come to appreciate the incredible complexity and beauty of the fungal world beneath our feet. Let this be an invitation to explore, discover, and, ultimately, protect the hidden world beneath us.

-Ravindra Goswami

Chapter 1

Introduction to Soil Fungi

Introduction to Soil Fungi

Soil fungi are among the most essential yet often overlooked components of terrestrial ecosystems. These microscopic organisms are found in almost every type of soil across the globe, from tropical rainforests to arid deserts, playing a pivotal role in the regulation of ecosystem processes. They are diverse, both in form and function, and are classified into several groups, including saprotrophic fungi, mycorrhizal fungi, parasitic fungi, and endophytic fungi. Saprotrophic fungi break down dead organic matter, contributing to nutrient cycling and soil structure, while mycorrhizal fungi form mutualistic relationships with plants, facilitating nutrient exchange. Parasitic fungi infect plants or other organisms, sometimes causing diseases, while endophytic fungi live within plant tissues without causing harm. This incredible diversity allows soil fungi to influence plant growth, soil health, and even atmospheric carbon levels, making them key players in both natural and managed ecosystems.

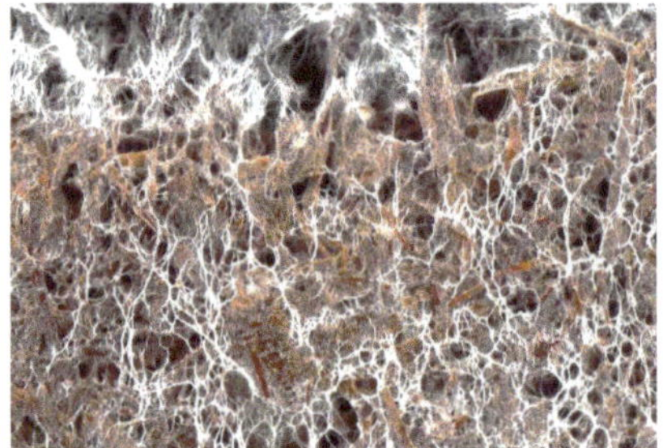

Fig 1A & B: Soil Fungi

The ecological roles of soil fungi are vast, beginning with their involvement in organic matter decomposition. Fungi are nature's recyclers, breaking down complex organic compounds like lignin and cellulose that other organisms cannot easily decompose. By doing so, they release essential nutrients such as nitrogen, phosphorus, and carbon back into the soil, making them available to plants and other microorganisms. This process, known as nutrient cycling, is critical for maintaining soil fertility and supporting plant life. Furthermore, soil fungi contribute to soil structure by binding soil particles together through their filamentous networks, or hyphae, which enhance water retention and soil aeration. This not only improves the physical properties of the soil but also makes it more resistant to erosion, thus playing a vital role in ecosystem sustainability.One of the most fascinating aspects of soil fungi is their symbiotic relationship with plants, particularly through the formation of mycorrhizal associations. In these partnerships, fungi colonize plant roots and extend their hyphal networks into the soil, effectively expanding the root system's reach. This allows plants to access water and nutrients, especially phosphorus, that would otherwise be out of reach. In exchange, plants provide the fungi with carbohydrates produced during photosynthesis. This mutualistic relationship is ancient, dating back hundreds of millions of years, and is crucial for the survival of the vast majority of plant species. In fact, nearly 90% of all land plants form some type of mycorrhizal association. By enhancing plant nutrient uptake, improving drought resistance, and even protecting against soil pathogens, mycorrhizal fungi are essential partners in the growth and health of both wild and cultivated plants.

Importance of Soil Fungi

Soil fungi are indispensable to the functioning of ecosystems, primarily because of their critical roles in nutrient cycling, organic matter decomposition, and maintaining soil structure. Fungi are among the few organisms capable of breaking down tough organic materials like lignin and cellulose found in plant debris. By decomposing these complex compounds, fungi release essential nutrients such as nitrogen, phosphorus, and carbon back into the soil, making them available to plants and other organisms. This process supports plant growth, sustains soil fertility, and underpins the entire food web, from microbes to larger fauna. Without fungi, organic matter would accumulate in ecosystems, leading to nutrient starvation for plants. Additionally, fungi's role in breaking down organic pollutants and sequestering carbon means they are also crucial in mitigating climate change. By storing carbon in the soil, fungi reduce the amount of carbon dioxide in the atmosphere, contributing to global carbon balance.

Moreover, soil fungi are fundamental to plant health through their symbiotic relationships, particularly in the form of mycorrhizae. Mycorrhizal fungi extend plant root systems and enhance their ability to absorb water and essential nutrients like phosphorus, improving plant growth and resistance to environmental stressors such as drought. These fungi also play a protective role by outcompeting or deterring soil pathogens, thus acting as a natural defense system for plants. In agriculture, mycorrhizal fungi are key to sustainable practices, reducing the need for synthetic fertilizers and enhancing crop productivity. Beyond agriculture, fungi contribute to ecosystem resilience by stabilizing soil and preventing erosion. Their hyphal networks physically bind soil particles together, improving soil structure and promoting aeration and water retention. In this way, soil fungi not only support plant life but also contribute to the long-term health and stability of entire ecosystems.

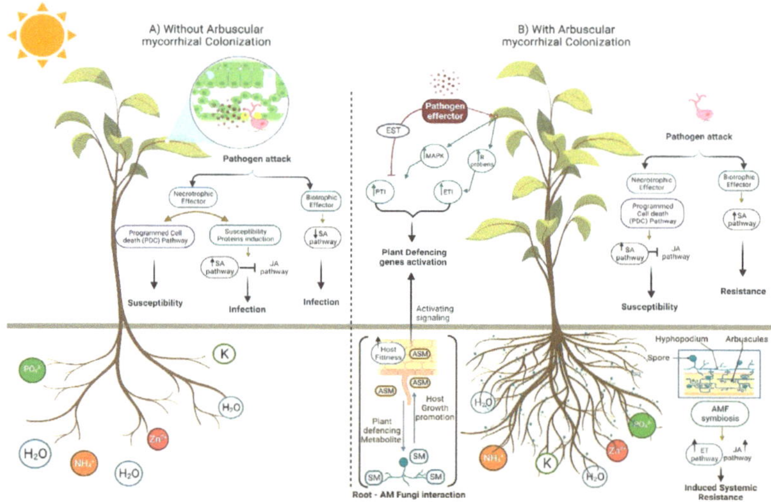

Fig 2: Hidden Power of Soil Fungi

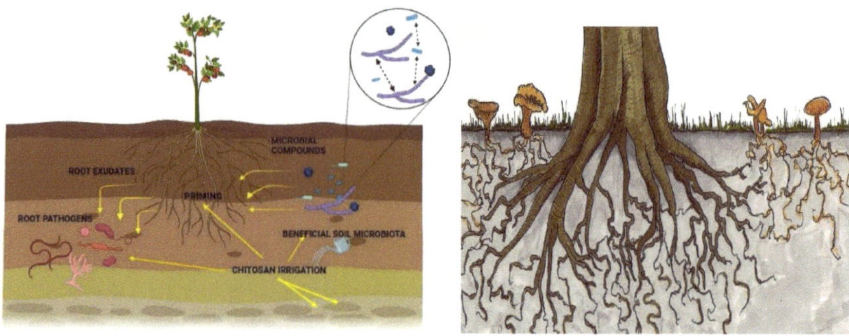

Fig 3: A. Soil Fungi & B. Mycorrhizal-fungi

1.1. Definition and classification of fungi.

The term **fungi** refer to a diverse kingdom of organisms that play crucial ecological roles, including decomposition, nutrient cycling, and symbiotic relationships with plants. Over the years, various mycologists and biologists have provided detailed definitions of fungi, highlighting their unique characteristics. Below are definitions of fungi by different authors, each emphasizing distinct aspects of fungal biology:

Alexopoulos, Mims, and Blackwell (1996), In their comprehensive textbook *"Introductory Mycology,"* Alexopoulos and his co-authors define fungi as: "Eukaryotic, heterotrophic organisms that lack chlorophyll and obtain their nutrients by absorption. Fungi reproduce by means of spores and have a vegetative body that is typically composed of hyphae, which form a mycelium. Their cell walls contain chitin, distinguishing them from plants."

Carl Woese (1977) In his pioneering work on the classification of organisms, Carl Woese introduced the idea of separating life into three domains (Bacteria, Archaea, Eukarya). In his classification, fungi fall under the domain Eukarya: "Fungi are eukaryotic microorganisms characterized by a distinct nuclear structure, whereby genetic material is enclosed within a membrane. Fungi differ from other eukaryotes in their mode of nutrition, which is absorptive rather than ingestive, and their ability to form extensive mycelial networks." Woese's definition places fungi within the larger context of eukaryotic organisms, highlighting their absorptive mode of nutrition and complex mycelial structures.

Deacon (2006), In his book *"Fungal Biology,"* John W. Deacon offers a more ecologically focused definition: "Fungi are heterotrophic organisms that play a pivotal role in ecosystems by breaking down complex organic materials, such as cellulose and lignin, and cycling essential nutrients. They are unique in their ability to form symbiotic relationships, such as mycorrhizae with plant roots, and can exist in a wide range of habitats, from soil to decaying wood." Deacon's definition underscores the ecological functions of fungi, particularly their role in decomposition and nutrient cycling, as well as their symbiotic relationships with plants.

Hawksworth (2001), David L. Hawksworth, a renowned mycologist, defined fungi in *"The Fungal Kingdom"* as: "Fungi are a kingdom of heterotrophic organisms that feed by absorption and produce a wide array of spores during their life cycles. They exhibit an extraordinary variety of forms, ranging from unicellular yeasts to complex multicellular structures like mushrooms and molds, with the majority existing as filamentous hyphae forming a mycelium." Hawksworth's definition is broad, encompassing the morphological diversity of fungi, from unicellular yeasts to multicellular forms, while also emphasizing their spore-producing capacity.

Carl Linnaeus (1753), In his early classification work, Linnaeus referred to fungi as one of the groups in the plant kingdom: "Fungi are plants that grow from decay, with a spongy structure, lacking chlorophyll and seeds." Though outdated by today's standards, Linnaeus's definition reflects the early perception of fungi as plants, though they were recognized even then as organisms that grow on decaying matter.

Kirk, Cannon, Minter, and Stalpers (2008), In *"Dictionary of the Fungi,"* Kirk et al. offer a concise and widely cited modern definition: "Fungi are heterotrophic eukaryotes, typically multicellular (except yeasts), which grow as filamentous hyphae forming a mycelium. They reproduce by sexual or asexual spores and obtain their nutrients through absorption. Fungi are distinct from other organisms by their cell walls, which contain chitin rather than cellulose." This definition emphasizes key features like heterotrophy, mycelial growth, and chitinous cell walls, distinguishing fungi from other life forms.

Margulis and Schwartz (1988), In their influential book *"Five Kingdoms: An Illustrated Guide to the Phyla of Life on Earth,"* Lynn Margulis and Karlene Schwartz offered the following definition: "Fungi are eukaryotic organisms that absorb dissolved organic material and form reproductive structures known as spores. Most fungi consist of long, threadlike structures called hyphae that aggregate to form a mycelium. Fungi are major decomposers in ecosystems and play a fundamental role in nutrient recycling." This definition highlights the absorptive mode of nutrition, hyphal structure, and ecological role of fungi as decomposers in ecosystems.

Carlene A. Peterson (2020), In *"Fungal Ecology: Principles and Practice,"* Carlene Peterson defines fungi from an ecological and functional perspective: "Fungi are eukaryotic organisms that occupy a unique niche in ecosystems, primarily as decomposers, but also as mutualists and pathogens. Their success is largely attributed to their ability to secrete extracellular enzymes that break down organic material and their formation of extensive hyphal networks which increase their absorptive surface area."

The **classification of fungi** is a detailed and evolving field that has undergone significant changes over the years due to advances in molecular biology and genetics. Fungi, traditionally classified based on their morphological characteristics and reproductive structures, are now categorized using molecular phylogenetics, which provides a more accurate understanding of their evolutionary relationships. Fungi are a distinct kingdom, separate from plants, animals, and protists, and their classification reflects their vast diversity in form, function, and habitat.

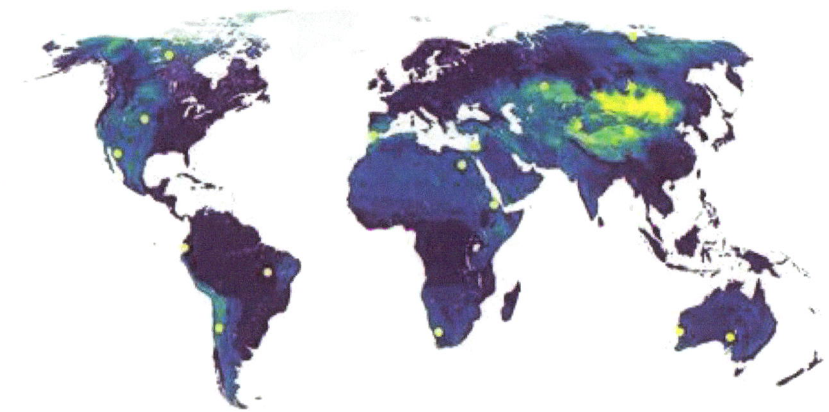

Fig 4: Fungal Biodiversity

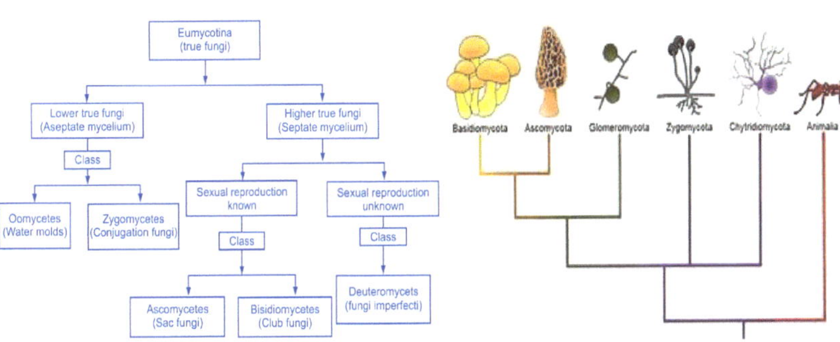

Fig 5: Classification of Fungi

Here's a detailed breakdown of the classification of fungi:

Kingdom: Fungi

The kingdom Fungi includes a wide variety of organisms, from microscopic molds and yeasts to large mushrooms. They are eukaryotic organisms, meaning their cells contain a nucleus, and they have chitin in their cell walls. Fungi are heterotrophic, obtaining their nutrients through absorption, and they reproduce by spores, which can be sexual or asexual.

Major Phyla of Fungi

Fungi are broadly classified into the following five major phyla, based on their sexual reproduction, structure, and genetic makeup:

1. Phylum Chytridiomycota (Chytrids)

Chytridiomycota, often referred to as chytrids, represent one of the most primitive and diverse phyla within the kingdom Fungi. They are unique among fungi due to their aquatic habitat and the presence of motile, flagellated spores called **zoospores,** which are capable of swimming through water. This flagellated structure is a key feature distinguishing chytrids from other fungi and is thought to be a trait inherited from their early evolutionary ancestors. Chytridiomycota include a variety of forms ranging from simple, single-celled organisms to complex, multicellular structures. They are found in diverse environments, including freshwater, soil, and the guts of herbivores, where they play crucial roles as decomposers, parasites, and mutualists. For example, some chytrids are saprobic, breaking down dead organic matter and contributing to nutrient cycling, while others form symbiotic relationships with plants, aiding in nutrient absorption.

A defining characteristic of Chytridiomycota is their reproductive cycle, which involves both asexual and sexual phases. Asexual reproduction occurs through the release of zoospores from a specialized structure known as a **sporangium**. These zoospores can swim to new locations, where they germinate and form new fungal colonies. Sexual reproduction in chytrids involves the fusion of gametes to form a **zygosporangium**, which develops into a thick-walled zygospore capable of surviving unfavorable conditions. Chytrids are notable for their ecological and medical significance; some species, like *Batrachochytrium dendrobatidis*, are pathogenic to amphibians, contributing to the global decline in amphibian populations. Despite their relatively simple structure, chytrids are vital for understanding the evolution of fungi and their ecological roles, serving as a living link to the early history of fungal development.

Characteristics:

- The **Chytridiomycota** are considered the most primitive fungi, with some of the earliest diverging lineages.
- They are primarily aquatic fungi, though some species are found in moist terrestrial environments.
- Unique among fungi, chytrids produce **flagellated zoospores** (motile spores), which enable them to swim in water.
- They typically have simple, often single-celled or filamentous structures and can be parasitic, saprobic (feeding on dead organic matter), or mutualistic.

Reproduction:

- Chytrids reproduce both sexually and asexually. Asexual reproduction occurs through the production of zoospores, while sexual reproduction involves the fusion of gametes.

Ecological Role:

- Chytrids play important roles in aquatic ecosystems as decomposers.
- Some chytrids are pathogens, such as *Batrachochytrium dendrobatidis*, which causes chytridiomycosis in amphibians, leading to widespread population declines.

2. Phylum Zygomycota (Zygomycetes)

Zygomycota, a phylum within the kingdom Fungi, comprises a diverse group of terrestrial fungi known for their distinctive reproductive structures and relatively simple morphology. Zygomycetes are characterized by their coenocytic hyphae, which are multinucleate and lack cross-walls (septa) dividing the cells. This structural feature allows for the continuous cytoplasmic flow within the hyphae. The most notable reproductive feature of Zygomycota is the formation of zygosporangia during sexual reproduction. When two compatible hyphae come together, they form a zygosporangium, where the nuclei fuse to produce a thick-walled zygosporangium containing the zygospore, a resilient, resting spore capable of withstanding harsh conditions. This zygospore eventually germinates to produce a new mycelium. Asexual reproduction occurs through the production of sporangia, which release numerous sporangiospores into the environment, allowing for rapid colonization of new substrates.

Zygomycota includes familiar genera such as *Rhizopus*, known for causing bread mold, and *Mucor*, which can be found on decaying organic matter. These fungi are saprobes, meaning they primarily decompose dead organic material, contributing significantly to nutrient cycling and soil fertility. Despite their ecological benefits, some Zygomycota species can be pathogenic to plants and animals. For instance, *Mucor* and other related fungi can cause mucormycosis, a serious infection in immunocompromised individuals. The phylum Zygomycota has been the subject of significant taxonomic revisions, with many species reclassified into newly established phyla such as Mucoromycota and Entomophthoromycota based on molecular phylogenetic data. These revisions reflect a more accurate understanding of fungal evolution and relationships, highlighting the dynamic nature of fungal taxonomy.

Characteristics:

- Zygomycota includes fungi that are mostly terrestrial, living in soil or on decaying plant or animal material.
- They have coenocytic hyphae, meaning their hyphae lack septa (cross-walls), making them multinucleate.
- Zygomycetes are fast-growing and commonly found on bread and fruit as molds, such as *Rhizopus stolonifer*, the black bread mold.

Reproduction:

- Zygomycota reproduce sexually by forming zygosporangia, a resistant structure where two different hyphal strands fuse. Inside the zygosporangium, a zygospore is produced, which can endure harsh conditions.
- Asexual reproduction occurs through sporangia that release sporangiospores.

Ecological Role:

- Zygomycetes are important decomposers of organic matter.
- Some are parasitic, while others form mutualistic relationships with plants, such as arbuscular mycorrhizae.

Recent molecular studies have led to the reclassification of many Zygomycetes into different groups, particularly Mucoromycota and Zoopagomycota, as Zygomycota is now considered polyphyletic (not representing a single evolutionary lineage).

3. Phylum Glomeromycota (Arbuscular Mycorrhizal Fungi)

Glomeromycota is a distinctive phylum of fungi primarily recognized for its crucial role in forming arbuscular mycorrhizae, a symbiotic relationship with the roots of most terrestrial plants. These fungi are integral to plant nutrition, enhancing the uptake of water and essential nutrients, particularly phosphorus, which is often limiting in soil. The hyphae of Glomeromycota extend into the plant root cells, creating highly branched structures known as arbuscules. These structures increase the surface area for nutrient exchange between the plant and the fungus. Additionally, Glomeromycota form large, multinucleate spores that are often produced underground, known as vesicular-arbuscular mycorrhizal spores. These spores are highly resistant to environmental conditions, aiding in the survival and proliferation of the fungi.

Despite their significant ecological role, Glomeromycota are obligate symbionts, meaning they cannot complete their life cycle without a host plant. They reproduce asexually, with spores germinating and establishing new mycorrhizal associations. Unlike many other fungi, Glomeromycota lack a known sexual reproductive phase. Their interaction with plant roots not only supports plant growth but also contributes to soil health by improving soil structure, water retention, and reducing erosion. This mycorrhizal association has been a key factor in the successful colonization of land by plants, highlighting the evolutionary significance of Glomeromycota in terrestrial ecosystems. The mutualistic relationship they maintain with plants underscores their importance in agriculture and forestry, where they are utilized to enhance crop yields and restore degraded lands.

Characteristics:

- Glomeromycota consists of fungi that form arbuscular mycorrhizae, a type of symbiotic relationship with the roots of most terrestrial plants.
- These fungi are critical for plant nutrition, as they enhance the plant's ability to absorb water and nutrients, particularly phosphorus, by extending their hyphal networks into the soil.
- They are obligate symbionts, meaning they cannot complete their life cycle without a plant host.

Reproduction:

- Glomeromycetes reproduce asexually, producing large, multinucleate spores underground. They do not have a known sexual reproductive phase.

Ecological Role:

- They are essential for the health of ecosystems as they play a major role in soil fertility and plant growth.
- By improving nutrient uptake and protecting plants from pathogens, they form a critical part of the plant-fungal symbiosis that has been present for hundreds of millions of years.

4. Phylum Ascomycota (Sac Fungi)

Ascomycota, also known as sac fungi, is one of the largest and most diverse phyla within the kingdom Fungi, encompassing a wide range of fungi from single-celled yeasts to complex multicellular forms like morels and truffles. This phylum is characterized by the production of sexual spores called ascospores, which are formed within sac-like structures known as asci. Asci are typically contained within larger fruiting bodies known as ascocarps. The defining reproductive feature of Ascomycota is the presence of these asci, where the ascospores are produced following the fusion of compatible hyphae. The diversity in morphology and habitat within Ascomycota is immense, with species ranging from the common bread mold *Neurospora* to the edible morel *Morchella*.

Asexual reproduction in Ascomycota is also significant and involves the production of conidia, which are asexual spores produced on specialized structures called conidiophores. Conidia are typically dispersed through the air, enabling rapid colonization of new environments. Ascomycota include numerous species with important ecological and economic roles. For example, yeasts such as *Saccharomyces cerevisiae* are essential in baking and brewing industries, while *Penicillium* species have been pivotal in the development of antibiotics like penicillin. Ascomycota also includes several plant pathogens, such as *Fusarium* and *Verticillium*, which cause significant crop diseases. This phylum's ability to adapt to various environments and its diverse reproductive strategies highlight its evolutionary success and its significant impact on both natural ecosystems and human activities.

Characteristics:

- Ascomycota is the largest phylum of fungi, including over 64,000 species. They are known as sac fungi because their sexual spores (ascospores) are produced in sac-like structures called asci.
- Ascomycetes include a wide range of fungi, such as yeasts (e.g., *Saccharomyces cerevisiae*), molds (e.g., *Penicillium*), and morels.
- They have septate hyphae, meaning their hyphae are divided into individual cells by cross-walls.

Reproduction:

- Ascomycota reproduce sexually by producing ascospores within asci, which are contained in larger fruiting bodies called ascocarps.
- Asexual reproduction occurs through the formation of conidia, which are non-motile spores produced on the surface of specialized structures called conidiophores.

Ecological Role:

- Ascomycetes play crucial roles as decomposers, breaking down organic material.

- Many are economically important, being used in food production (yeasts in fermentation), pharmaceuticals (antibiotics like penicillin), and biotechnology.
- Some Ascomycota species are pathogens of plants (e.g., *Claviceps purpurea*, which causes ergot on rye) and animals, including humans.

5. Phylum Basidiomycota (Club Fungi)

Basidiomycota, commonly known as club fungi, is a diverse and ecologically significant phylum within the kingdom Fungi. It includes a wide array of fungi, from the familiar mushrooms and puffballs to rusts and smuts that affect plants. The defining feature of Basidiomycota is the production of sexual spores called basidiospores, which are formed on specialized structures known as basidia. Basidia are typically found on the surface of fruiting bodies called basidiocarps. The process of sexual reproduction involves the fusion of hyphae from two compatible mating types, leading to the formation of basidia where meiosis occurs to produce the basidiospores. These spores are then dispersed into the environment to germinate and form new fungal mycelium. The intricate structure of basidiocarps, which includes familiar mushrooms with their gills or pores, plays a crucial role in spore dispersal and reproduction.

Basidiomycota exhibits considerable ecological diversity and impact. Many members, such as *Agaricus bisporus* (the common mushroom) and *Boletus edulis* (porcini), are edible and economically important for culinary use. Additionally, Basidiomycota includes fungi that form ectomycorrhizae, symbiotic associations with plant roots that enhance nutrient uptake and contribute to soil health. These relationships are crucial for the growth of many forest trees and plants. On the other hand, some Basidiomycota species are significant plant pathogens, such as the rust fungi (*Puccinia*) and smut fungi (*Ustilago*), which can cause severe crop losses. The ability of Basidiomycota to decompose complex organic materials, such as lignin in wood, further underscores their ecological role in nutrient cycling and forest ecosystem dynamics. Their diverse reproductive strategies and ecological roles highlight their evolutionary success and importance in various habitats.

Characteristics:

- Basidiomycota, known as club fungi, include some of the most familiar fungi, such as mushrooms, puffballs, shelf fungi, and rusts and smuts (plant pathogens).
- The characteristic reproductive structure of this group is the basidium, a microscopic club-shaped structure where sexual spores (basidiospores) are produced.
- They have septate hyphae with complex dikaryotic (two nuclei per cell) life stages.

Reproduction:

- Basidiomycetes reproduce sexually through the formation of basidiospores on basidia, which are located on fruiting bodies called basidiocarps (e.g., mushrooms).

- Asexual reproduction is less common but can occur through the formation of conidia or other spore types.

Ecological Role:

- Basidiomycetes are important decomposers of wood and leaf litter, especially because of their ability to break down lignin, a complex component of plant cell walls.
- They form mutualistic relationships with plants in the form of ectomycorrhizae, essential for forest ecosystems.
- Some are significant pathogens in agriculture, such as rusts and smuts that affect cereal crops.

Fungal Imperfections: Deuteromycota (Imperfect Fungi)

Deuteromycota, commonly referred to as imperfect fungi, is a group of fungi classified not by their evolutionary lineage but by the absence of a known sexual reproduction phase. Historically, fungi have been classified based on their sexual reproductive structures, but for Deuteromycota, no sexual stage has been observed. This led to their classification as "imperfect," as opposed to fungi that reproduce sexually, known as "perfect fungi." Deuteromycetes reproduce asexually through the production of conidia, asexual spores formed on structures called conidiophores. Despite this limitation in classification, many Deuteromycetes are vital in natural ecosystems and human applications. Some play crucial roles in the decomposition of organic matter, while others are significant pathogens of plants, animals, and even humans. However, some Deuteromycetes, such as *Penicillium* and *Aspergillus*, are of immense importance in medicine and industry. *Penicillium* species are the source of the antibiotic penicillin, while *Aspergillus* is involved in producing industrial enzymes and fermenting foods like soy sauce.

Advances in molecular biology have allowed scientists to reclassify many Deuteromycota into the Ascomycota and Basidiomycota phyla as their sexual stages become identified. With DNA sequencing, researchers now understand that Deuteromycota is not a monophyletic group, meaning it does not stem from a single common ancestor but represents fungi from various lineages lacking observed sexual reproduction. As molecular techniques continue to evolve, many species traditionally placed in Deuteromycota are now integrated into other fungal phyla based on genetic similarities. This reclassification reflects our growing understanding of fungal evolution and genetics, but the term Deuteromycota is still sometimes used for convenience in referring to fungi without a known sexual stage, particularly in medical and industrial contexts where asexual reproduction predominates.

Characteristics:

Deuteromycota is not a true phylum but a classification used for fungi whose sexual stage is unknown or has not been observed.

These fungi reproduce only asexually, typically by producing conidia.

Once the sexual stage is discovered, many of these fungi are reclassified into either Ascomycota or Basidiomycota.

Examples include species of *Penicillium* and *Aspergillus*, which are widely used in industry and research but were initially placed in this group due to the lack of knowledge about their sexual reproduction.

Summary of Fungal Classification

- **Chytridiomycota**: Primitive, aquatic fungi with flagellated spores.
- **Zygomycota (now largely split into Mucoromycota and Zoopagomycota)**: Terrestrial fungi known for producing resistant zygospores.
- **Glomeromycota**: Arbuscular mycorrhizal fungi, forming mutualistic relationships with plant roots.
- **Ascomycota**: Sac fungi, producing spores in asci; includes yeasts, molds, and morels.
- **Basidiomycota**: Club fungi, producing spores on basidia; includes mushrooms, puffballs, and plant pathogens.
- **Deuteromycota**: Imperfect fungi, where sexual reproduction is unknown.

The classification of fungi continues to evolve as new molecular data emerges, refining our understanding of their evolutionary relationships and ecological roles.

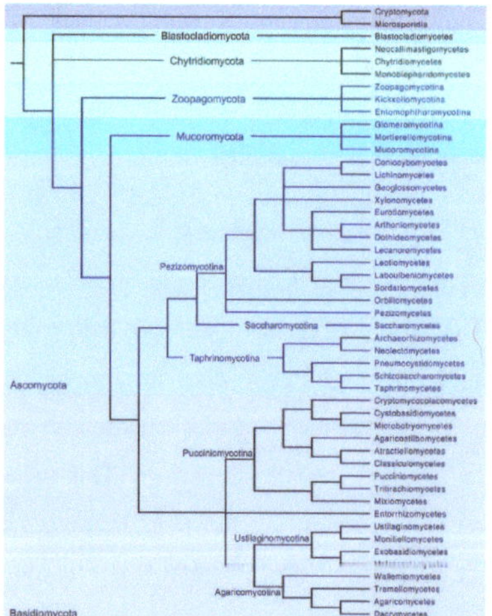

Fig 6: Fungal Tree: Depicted Classification

1.2.Overview of fungi in the ecosystem, with a focus on soil environments.

Fungi play a crucial and multifaceted role in ecosystems, particularly within soil environments, where they significantly contribute to nutrient cycling, soil structure, and overall ecosystem health. In soil, fungi are pivotal decomposers, breaking down complex organic matter such as dead plant material, leaf litter, and woody debris. This decomposition process releases essential nutrients like nitrogen, phosphorus, and potassium back into the soil, making them available for plant uptake. Fungi decompose organic material through the secretion of extracellular enzymes that break down complex compounds, including lignin and cellulose, which are otherwise resistant to decomposition. This decomposition process not only recycles nutrients but also contributes to the formation of humus, enhancing soil fertility and structure. By forming extensive mycelial networks, fungi can improve soil aeration and water retention, thereby promoting healthy root growth and stability.

In addition to their role in decomposition, fungi form symbiotic relationships with plants through structures such as mycorrhizae. Mycorrhizal fungi, particularly those in the phyla Glomeromycota and Basidiomycota, associate with plant roots to form mutualistic relationships. These fungi extend their hyphal networks into the soil, increasing the surface area for nutrient absorption, particularly phosphorus, which is often limited in soil. In return, plants provide the fungi with organic carbon derived from photosynthesis. This symbiosis is essential for the health of many terrestrial ecosystems, supporting plant growth and resilience. Furthermore, soil fungi play a role in suppressing plant pathogens and maintaining soil biodiversity. Through their interactions with various soil organisms and their contributions to soil processes, fungi are integral to sustaining the health and functionality of soil ecosystems, highlighting their importance in maintaining ecological balance and productivity.

Fungi are integral to ecosystem functioning, especially in soil environments, where they perform vital roles in nutrient cycling, soil formation, and plant health. In soil ecosystems, fungi are primarily involved in the decomposition of organic matter. They break down complex plant materials such as lignin, cellulose, and chitin through the action of extracellular enzymes. This decomposition process converts organic matter into simpler compounds, which are then mineralized into essential nutrients like nitrogen, phosphorus, and potassium. These nutrients are critical for plant growth and contribute to soil fertility. The activity of decomposer fungi also helps in the formation of humus, which improves soil structure, enhances water retention, and provides a stable environment for plant roots. The extensive mycelial networks of fungi increase soil aeration and facilitate the movement of water and nutrients through the soil matrix, further benefiting plant health.

Beyond decomposition, fungi form symbiotic relationships with plants through **mycorrhizae**, which are mutualistic associations between plant roots and fungi. Mycorrhizal fungi, including those from the phyla

Glomeromycota, Ascomycota, and Basidiomycota, extend their hyphal networks into the soil, enhancing the plant's ability to absorb water and essential nutrients, particularly phosphorus, which is often scarce in soils. In return, plants provide the fungi with organic carbon derived from photosynthesis. This relationship is crucial for the growth and survival of many plant species and contributes to the overall health and productivity of terrestrial ecosystems. Additionally, soil fungi play a role in suppressing soil-borne pathogens and maintaining soil biodiversity by interacting with other soil microorganisms and contributing to the ecological balance. Their multifaceted functions highlight their importance in sustaining soil health and ecosystem stability.

Fungi play a critical and diverse role in ecosystems, influencing nutrient cycling, plant health, and community dynamics across various environments. Their ecological functions are wide-ranging and essential for the maintenance of ecosystem stability and productivity.

1. Nutrient Cycling and Decomposition: Fungi are primary decomposers in most ecosystems. They break down complex organic materials, such as dead plants, leaves, wood, and animal remains, using extracellular enzymes to degrade lignin, cellulose, and chitin. This decomposition process releases nutrients such as nitrogen, phosphorus, and potassium back into the soil, which are vital for plant growth. By transforming organic matter into humus, fungi enhance soil structure and fertility. Their role in nutrient cycling is crucial for the sustainability of both terrestrial and aquatic ecosystems.

2. Symbiotic Relationships: Many fungi form mutualistic relationships with plants through mycorrhizae, which are symbiotic associations between fungal hyphae and plant roots. Mycorrhizal fungi extend their hyphal networks into the soil, increasing the surface area for nutrient and water absorption, particularly for nutrients that are less available, such as phosphorus. In return, plants supply fungi with organic carbon derived from photosynthesis. This mutualistic relationship is vital for the growth and health of many plant species and can significantly enhance soil fertility and plant productivity.

3. Soil Health and Structure: Fungal mycelia contribute to the physical properties of soil. The hyphal networks formed by fungi improve soil aggregation and structure by binding soil particles together, which enhances soil porosity and water infiltration. This improved soil structure reduces erosion, increases water-holding capacity, and supports healthier plant roots. Additionally, fungi contribute to the formation of soil aggregates, which are important for soil stability and aeration.

4. Pathogens and Disease: Fungi can also act as pathogens in ecosystems. Plant pathogenic fungi, such as rusts, smuts, and molds, can cause diseases that affect crop yields and natural vegetation. However, these interactions are part of natural ecosystems and contribute to ecological balance by regulating plant populations and influencing plant community dynamics. Some fungi are also pathogenic to animals and humans, causing diseases that can have significant health impacts.

5. Biodiversity and Ecosystem Dynamics: Fungi contribute to ecosystem biodiversity by occupying various ecological niches and interacting with a wide range of organisms. They form complex relationships with bacteria, insects, and other soil microorganisms, influencing microbial communities and ecosystem processes. Fungal diversity and interactions help maintain ecological balance and resilience in response to environmental changes.

6. Bioremediation and Environmental Restoration: Certain fungi have the ability to degrade pollutants and toxins, a process known as bioremediation. Fungi can break down complex organic pollutants, such as pesticides and hydrocarbons, into less harmful substances. This capability makes them valuable for environmental cleanup and restoration efforts, particularly in contaminated soils and water bodies.

Overall, fungi are essential components of ecosystems, playing multifaceted roles that support nutrient cycling, soil health, plant productivity, and overall ecological balance. Their diverse functions and interactions highlight their importance in maintaining the health and stability of natural environments.

Fungi are fundamental to soil environments, playing crucial roles in nutrient cycling, soil health, and plant productivity. Their presence and activity significantly influence the ecological dynamics and functionality of soils.

1. Decomposition and Nutrient Cycling: One of the primary roles of fungi in soil environments is decomposition. Fungi decompose complex organic matter such as dead plants, leaf litter, and woody debris. They achieve this through the secretion of extracellular enzymes that break down substances like lignin, cellulose, and chitin. This decomposition process releases essential nutrients such as nitrogen, phosphorus, and potassium back into the soil, making them available for plant uptake. By converting organic matter into simpler compounds, fungi contribute to the formation of humus, which enhances soil fertility, structure, and water-holding capacity. This nutrient cycling is critical for maintaining soil health and supporting plant growth.

2. Mycorrhizal Associations: Fungi form symbiotic relationships with plant roots through structures known as mycorrhizae. Mycorrhizal fungi, including those from the phyla Glomeromycota, Basidiomycota, and Ascomycota, extend their hyphal networks into the soil and plant roots, significantly increasing the surface area for nutrient and water absorption. This interaction is particularly important for the uptake of nutrients like phosphorus, which is often limited in soil. In return, plants provide the fungi with organic carbon from photosynthesis. This mutualistic relationship enhances plant growth, improves soil structure, and boosts soil fertility. Mycorrhizae are vital for the health of many terrestrial ecosystems, including forests, grasslands, and agricultural systems.

3. Soil Structure and Health: Fungal mycelium contributes to the physical properties of soil. The extensive networks of fungal hyphae help bind soil particles together, forming aggregates that improve soil structure and stability. This aggregation enhances soil porosity, water infiltration, and aeration, which are crucial for healthy plant root development and soil erosion prevention. By improving soil structure, fungi also increase the soil's water-holding capacity and reduce runoff, contributing to better drought resistance and overall soil health.

4. Pathogenic Interactions: Fungi can also act as pathogens in soil environments, affecting plants, animals, and other fungi. Plant pathogenic fungi, such as rusts, smuts, and molds, can cause diseases that impact crop yields and natural vegetation. While these pathogens can negatively affect plant health and productivity, they also play a role in natural ecosystems by regulating plant populations and influencing community dynamics. Additionally, soil-borne fungal pathogens can affect soil microbiomes and nutrient cycling processes, further impacting ecosystem functions.

5. Soil Biodiversity and Interactions: Fungi are a critical component of soil biodiversity, interacting with a variety of microorganisms, including bacteria, protists, and other fungi. These interactions contribute to the complexity and stability of soil ecosystems. Fungi can influence microbial community composition and activity, impacting nutrient cycling, decomposition rates, and soil health. Their diverse interactions with other soil organisms help maintain ecological balance and resilience in response to environmental changes.

6. Bioremediation and Environmental Management: Some fungi have the ability to degrade pollutants and toxins, a process known as bioremediation. Fungi can break down organic pollutants such as pesticides, heavy metals, and hydrocarbons into less harmful substances. This capability makes them valuable for soil and water contamination cleanup and environmental restoration efforts. By utilizing fungi in bioremediation, it is possible to mitigate pollution and restore soil health in contaminated sites.

Fungi are essential to soil environments, performing diverse and critical functions that support nutrient cycling, soil health, plant productivity, and overall ecosystem stability. Their roles in decomposition, mycorrhizal symbiosis, soil structure, and bioremediation underscore their importance in maintaining healthy and productive soil ecosystems.

1.3. Brief history of fungal research and its importance in ecology

The history of fungal research is a fascinating journey through the discovery and understanding of one of nature's most versatile and ecologically significant groups of organisms. From early observations to modern scientific advancements, the study of fungi has profoundly impacted our understanding of ecosystems and their functioning.

Early Observations and Classification (Pre-19th Century): Early knowledge of fungi dates back to ancient civilizations, where mushrooms and other fungi were noted for their medicinal and culinary uses. Ancient Greek and Roman scholars, such as Theophrastus and Pliny the Elder, made some of the earliest documented observations of fungi, though their biological significance was not fully understood. In the Middle Ages, the use of fungi in traditional medicine continued, but systematic scientific study was limited.

Foundations of Mycology (19th Century): The formal study of fungi began to take shape in the 19th century with the advent of mycology as a distinct scientific discipline. The work of early mycologists such as Elias Magnus Fries and Louis René Tulasne laid the groundwork for fungal taxonomy and classification. Fries, often referred to as the "father of mycology," developed a system for classifying fungi based on their reproductive structures, which became the basis for future fungal classification. Tulasne's detailed studies of fungal reproduction further advanced the understanding of fungal life cycles and structures.

Advancements and Discoveries (20th Century): The 20th century brought significant advancements in fungal research, including the development of modern taxonomic systems and an increased focus on the ecological roles of fungi. The discovery of the antibiotic properties of penicillin by Alexander Fleming in 1928 marked a milestone in medical mycology, showcasing the practical importance of fungi in medicine. This discovery led to the development of antibiotics and revolutionized the treatment of bacterial infections. The 20th century also saw the emergence of mycorrhizal research, highlighting the critical role of fungi in forming symbiotic relationships with plants and enhancing nutrient uptake.

Modern Mycology and Ecological Importance (21st Century): In recent decades, fungal research has expanded to include molecular and genetic studies, providing deeper insights into fungal biology, evolution, and ecology. Advances in DNA sequencing and phylogenetics have led to the reclassification of many fungi and a better understanding of their evolutionary relationships. Modern mycology has emphasized the ecological importance of fungi, including their roles in nutrient cycling, soil health, and plant interactions. Fungi are now recognized for their critical contributions to ecosystem functioning, including decomposition, symbiotic relationships with plants (mycorrhizae), and bioremediation of pollutants.

Importance in Ecology: Fungal research has underscored the vital roles fungi play in ecosystems. Their ability to decompose organic matter and recycle nutrients is fundamental to soil health and fertility. Mycorrhizal fungi form essential symbiotic relationships with plants, supporting plant growth and ecosystem productivity. Fungi also contribute to soil structure and stability, enhance water retention, and support biodiversity through their interactions with other soil microorganisms. Additionally, fungi play a role in bioremediation, aiding in the cleanup of contaminated environments and contributing to environmental sustainability.

The history of fungal research reflects a growing appreciation of fungi's diverse roles and ecological significance. From early observations to advanced molecular studies, the field of mycology has revealed the essential contributions of fungi to ecosystem health, nutrient cycling, and environmental management. The continued exploration of fungal biology and ecology promises to uncover further insights into their complex interactions and impacts on the natural world.

1.4.Conclusion

soil fungi represent a critical component of terrestrial ecosystems, playing indispensable roles in nutrient cycling, plant health, and soil structure. Throughout this book, we have explored the incredible diversity of soil fungi, from decomposers that break down complex organic matter to mycorrhizal fungi that form symbiotic relationships with plants, ensuring the efficient uptake of essential nutrients like phosphorus. Their role in transforming organic matter into simpler compounds like humus not only enriches the soil but also supports an intricate web of life that sustains both plant and microbial communities. Fungi's ability to break down substances that other organisms cannot, like lignin and cellulose, highlights their evolutionary adaptation and ecological significance. Without fungi, the organic matter in ecosystems would accumulate, depriving plants of the nutrients they need, ultimately leading to ecosystem collapse.

The ecological services provided by soil fungi go beyond nutrient recycling; they play an active role in enhancing soil structure, increasing its water-holding capacity, and preventing erosion through their extensive hyphal networks. Fungi are also key players in environmental resilience, improving plant tolerance to environmental stresses such as drought and pathogens. The symbiotic relationships between fungi and plants, particularly through mycorrhizae, demonstrate the interconnectedness of life in the soil and the importance of fungi for ecosystem stability and agricultural sustainability. As we look to the future, the study of soil fungi will continue to offer insights into sustainable land management, climate resilience, and even bioremediation efforts to clean contaminated environments. This exploration of the hidden world beneath our feet underscores the vital role soil fungi play in maintaining the balance and health of our planet's ecosystem

Chapter 2

Fungal Biodiversity in Soil

Fungal biodiversity in soil is a critical component of terrestrial ecosystems, contributing to soil health, nutrient cycling, plant growth, and overall ecosystem stability. Soil is home to an astonishingly diverse community of fungi, which perform a wide array of ecological functions. Fungal species in soil include saprotrophic fungi, mycorrhizal fungi, parasitic fungi, and endophytic fungi, each playing unique roles in sustaining the ecosystem. Saprotrophic fungi are key decomposers, breaking down organic matter such as dead plant and animal material, facilitating the release of essential nutrients like nitrogen, phosphorus, and carbon back into the soil. This decomposition process not only recycles nutrients but also improves soil structure by forming humus, a stable organic material that enhances water retention and aeration, and promotes healthy plant root growth. Mycorrhizal fungi, particularly arbuscular mycorrhizal fungi (AMF) and ectomycorrhizal fungi, form mutualistic associations with plant roots, where they exchange nutrients for carbohydrates. These fungi significantly enhance plant nutrient uptake, especially phosphorus and nitrogen, by extending the root's reach into the soil through an intricate network of fungal hyphae. Mycorrhizal associations are widespread and essential for the survival of many plant species, particularly in nutrient-poor soils, and are crucial for forest and grassland ecosystems. Ectomycorrhizal fungi, in particular, form relationships with many tree species in temperate and boreal forests, playing a critical role in forest ecology by aiding in tree nutrient acquisition and improving resilience to environmental stressors like drought and disease. Parasitic fungi, while often considered detrimental due to their ability to cause diseases in plants and other organisms, also contribute to soil biodiversity by influencing the dynamics of plant and microbial communities. These fungi can regulate plant populations by infecting weaker individuals, thereby allowing for more diverse plant communities to thrive. Endophytic fungi, which live inside plant tissues without causing harm, are also abundant in soil ecosystems. They often enhance plant growth by increasing stress tolerance and protecting plants from pathogens.

Fungal biodiversity in soil is also influenced by various environmental factors, including soil pH, temperature, moisture, organic matter content, and vegetation type. Fungi tend to thrive in moist, well-aerated soils rich in organic matter, but some species are adapted to more extreme conditions, such as arid soils, acidic environments, or nutrient-poor habitats. The diversity of fungi in soil is highly dynamic and can vary greatly across different ecosystems, from tropical rainforests, which typically have high fungal diversity, to deserts, where specialized fungi adapt to extreme conditions.

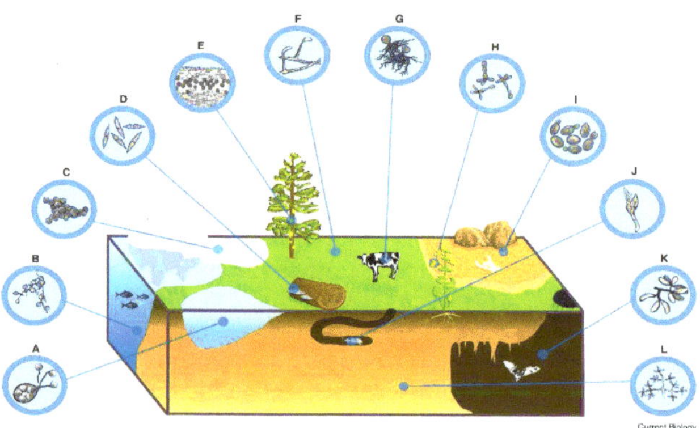

Fig 8: Fungal Biodiversity

The presence and diversity of fungi are also influenced by human activities such as agriculture, deforestation, and pollution. Intensive farming practices that involve the excessive use of fertilizers and pesticides can reduce fungal diversity by altering soil chemistry and eliminating sensitive species. On the other hand, conservation practices, such as organic farming, reforestation, and the use of biofertilizers, can promote fungal biodiversity by creating more favorable conditions for fungal growth and maintaining natural soil processes.

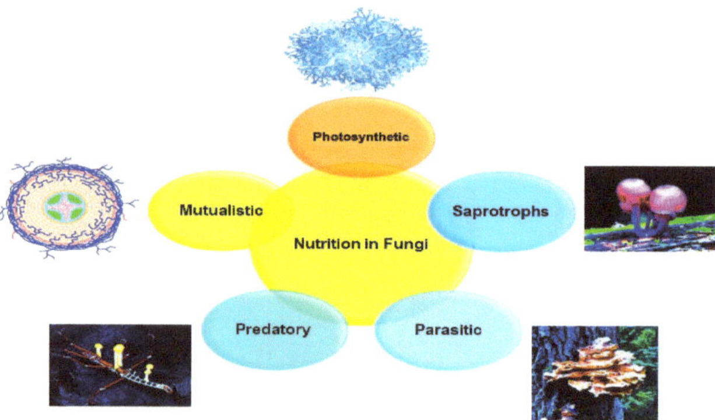

Fig 8: Types of soil fungi

The importance of maintaining fungal biodiversity in soil cannot be overstated, as it underpins the ecological processes that drive ecosystem productivity and resilience. Fungi contribute to carbon sequestration, soil fertility, and plant health, and they are integral to the structure of soil food webs. Without a diverse and functioning fungal community, soils would lose their ability to support plant life and cycle nutrients

25

effectively, leading to decreased biodiversity and ecosystem degradation. Therefore, the preservation of fungal biodiversity in soil is essential for maintaining healthy ecosystems and mitigating the impacts of environmental change.

2.1. Types of soil fungi: Saprotrophic, mycorrhizal, parasitic, and endophytic fungi.

Soil fungi can be broadly categorized into three main types based on their ecological roles: decomposers (saprotrophic fungi), mutualists (mycorrhizal fungi), and pathogens (parasitic fungi).

Decomposers are among the most crucial for ecosystem functioning, as they break down complex organic materials like dead plant matter, leaves, and wood, returning essential nutrients such as nitrogen, phosphorus, and carbon to the soil. These fungi, including species from the phyla Ascomycota and Basidiomycota, secrete powerful enzymes that decompose tough materials like lignin and cellulose, which are difficult for other organisms to break down. Saprotrophic fungi are fundamental in the formation of humus, which enhances soil fertility, structure, and water retention. They are the driving force behind the continuous cycling of organic matter in soils, which supports plant growth and sustains the microbial community.

Mutualistic fungi, particularly mycorrhizal fungi, form symbiotic relationships with plant roots, playing a critical role in nutrient exchange. Ectomycorrhizal fungi (mainly from Basidiomycota and Ascomycota) form a sheath around plant roots, while arbuscular mycorrhizal fungi (from the phylum Glomeromycota) penetrate root cells to form intricate structures. In these relationships, fungi extend their hyphal networks deep into the soil, dramatically increasing the surface area for nutrient and water absorption, especially for scarce nutrients like phosphorus. In return, the plant provides the fungi with carbon in the form of sugars from photosynthesis. This mutualistic relationship enhances plant growth, improves soil structure, and contributes to the long-term stability of ecosystems. Lastly, pathogenic fungi act as parasites, causing diseases in plants by infecting their roots or other tissues, which can lead to significant agricultural losses. While they can be harmful, these fungi also play a role in regulating plant populations and maintaining ecological balance. Together, these different types of soil fungi are essential for maintaining the dynamic balance and health of soil ecosystems.

Saprotrophic fungi

Saprotrophic fungi are a diverse group of fungi that play a fundamental role in ecosystems as primary decomposers. Their ecological function is crucial because they break down dead organic matter, recycling nutrients back into the soil. These fungi use extracellular enzymes to decompose complex organic materials such as lignin, cellulose, chitin, and other components of dead plants, animals, and microorganisms. Through this decomposition process, saprotrophic fungi convert organic material into simpler compounds, making nutrients like nitrogen, phosphorus, and carbon available for uptake by plants and other organisms. This recycling of nutrients is essential for soil fertility and ecosystem productivity, especially in forested and grassland ecosystems where large amounts of organic matter accumulate.

Saprotrophic fungi belong to several phyla, including Basidiomycota, Ascomycota, and Zygomycota. Common examples include mushrooms, molds, and yeasts, with some well-known species like Agaricus bisporus (the common mushroom) and Penicillium species. The presence of saprotrophic fungi is often associated with the formation of humus, a vital component of soil that enhances water retention, aeration, and nutrient-holding capacity. Fungi like Trichoderma and Aspergillus species are also used industrially in the production of enzymes and antibiotics due to their powerful breakdown capabilities. In addition to nutrient recycling, saprotrophic fungi play a role in soil formation by breaking down organic matter into humic substances, which improve soil structure and provide a stable environment for microbial and plant communities. Their ability to degrade tough, resistant materials like wood makes them invaluable in maintaining the balance of ecosystems by preventing the accumulation of organic debris.

Mycorrhizal fungi

Mycorrhizal fungi form essential symbiotic relationships with the roots of most terrestrial plants, greatly enhancing nutrient and water uptake while receiving carbohydrates and other organic compounds in return. These mutualistic associations are fundamental to plant health and ecosystem productivity, as they extend the effective root surface area far beyond the plant's natural reach. The two primary types of mycorrhizae are ectomycorrhizae and arbuscular mycorrhizae (AM). Ectomycorrhizal fungi (primarily from the Basidiomycota and Ascomycota phyla) form a sheath around the outside of plant roots and penetrate the root cortex, facilitating nutrient exchange between the fungi and the plant. These fungi are particularly important for woody plants, including many trees in temperate and boreal forests, helping them to absorb nutrients like nitrogen and phosphorus from the soil. On the other hand, arbuscular mycorrhizal fungi (from the Glomeromycota phylum) penetrate the cells of plant roots, forming intricate, tree-like structures known as arbuscules within the root cells. These structures enable the fungi to efficiently transfer phosphorus and other nutrients to the plant, while the fungi gain access to sugars produced through photosynthesis.

The benefits of mycorrhizal associations extend beyond nutrient exchange. Mycorrhizal fungi significantly improve plant tolerance to environmental stresses, such as drought, soil compaction, and pathogens. By improving the physical properties of soil, these fungi enhance soil structure through the formation of aggregates, which improve aeration, water retention, and root penetration. In agricultural systems, mycorrhizae contribute to sustainable farming practices by reducing the need for chemical fertilizers and enhancing crop resilience. Moreover, mycorrhizal networks, often referred to as the "wood wide web," allow plants to communicate and share resources through interconnected fungal hyphae. This underground

27

network can facilitate the transfer of nutrients between plants, support seedling establishment, and enhance biodiversity by stabilizing ecosystems. Thus, mycorrhizal fungi are not only essential for individual plant health but also play a crucial role in maintaining the structure and stability of entire ecosystems.

Parasitic fungi

Parasitic fungi are a diverse group of fungi that obtain nutrients by living on or inside other organisms (hosts), often causing harm in the process. Unlike saprotrophic fungi, which decompose dead organic matter, parasitic fungi actively invade living hosts, such as plants, animals, insects, or other fungi, to derive nourishment. Parasitic fungi can cause diseases that severely impact agricultural crops, forests, and even human health. Their feeding mechanism involves penetrating the host's tissues using specialized structures, such as haustoria or appressoria, which allow them to absorb nutrients directly from the host cells without immediately killing them. This long-term parasitic relationship can weaken the host, cause disease symptoms, and sometimes result in death.

In plants, parasitic fungi can lead to devastating diseases that affect global food production and ecosystems. Notable examples include rusts and smuts, which infect cereals and other crops. For instance, Puccinia graminis, the fungus responsible for stem rust in wheat, is a major agricultural pathogen that can lead to significant yield losses. Another notorious example is Phytophthora infestans, the oomycete responsible for the Irish potato famine. These fungi often attack the vascular systems of plants, impairing nutrient and water transport, leading to stunted growth, wilting, and eventual death. In forests, parasitic fungi such as Armillaria species cause root rot in trees, destabilizing entire ecosystems by killing large numbers of trees over time.

In the animal kingdom, parasitic fungi also pose serious threats. One prominent example is Batrachochytrium dendrobatids, the chytrid fungus responsible for the decline and extinction of many amphibian species worldwide. Parasitic fungi also infect insects; for instance, Cordyceps species are known for their dramatic infections of insect hosts, controlling their behavior and ultimately killing them to spread spores. In humans and other animals, fungi like Candida albicans can become opportunistic pathogens, causing infections when the host's immune system is weakened. The impact of parasitic fungi extends beyond individual organisms, as they can alter population dynamics, ecosystem balance, and biodiversity. However, while they are often seen as destructive, parasitic fungi also play important roles in regulating populations and maintaining ecological balance by controlling host densities and influencing competition among species.

Endophytic fungi

Endophytic fungi are a fascinating group of fungi that live inside plant tissues without causing apparent harm to their hosts. Unlike parasitic fungi, which extract nutrients at the expense of the host's health, endophytic fungi form a symbiotic or neutral relationship with plants, often conferring various ecological benefits. These fungi colonize the internal spaces of plant tissues, such as leaves, stems, and roots, and coexist with their host for most or all of the plant's life cycle. In return, the plant provides the fungi with shelter and access to nutrients. Endophytic fungi are incredibly diverse, representing various fungal taxa, including Ascomycota, Basidiomycota, and even some groups within Zygomycota. They have been found in almost every plant species examined, from grasses and trees to agricultural crops and medicinal plants.

The relationship between endophytic fungi and their plant hosts is often mutualistic, meaning that both parties benefit. One of the primary advantages conferred by endophytic fungi to plants is increased resistance to biotic and abiotic stresses. Some endophytes produce secondary metabolites that deter herbivores and inhibit the growth of plant pathogens, providing the plant with a natural defense system. For instance, certain endophytic fungi produce alkaloids and other toxic compounds that protect grasses from being eaten by herbivores. Additionally, endophytic fungi can enhance the host plant's tolerance to environmental stressors such as drought, salinity, and extreme temperatures. This is achieved through various mechanisms, including the production of hormones that promote plant growth, the enhancement of nutrient uptake (particularly phosphorus and nitrogen), and improved water retention in plant tissues. In agricultural contexts, this relationship can be particularly beneficial, as crops colonized by beneficial endophytes exhibit greater resilience to diseases and environmental challenges, potentially reducing the need for chemical inputs such as pesticides and fertilizers.

Research on endophytic fungi is expanding rapidly, revealing their potential applications in biotechnology, agriculture, and medicine. Some endophytic fungi have been found to produce bioactive compounds with antimicrobial, anticancer, and antiviral properties, highlighting their potential for pharmaceutical use. In agriculture, the use of endophytes as biological control agents or plant growth promoters is being explored as a sustainable alternative to chemical-based farming practices. Overall, endophytic fungi play a critical, often overlooked role in plant health and productivity, acting as invisible allies that help plants adapt and thrive in diverse environments.

2.2.Examples of important soil fungi species.

Several species of soil fungi play critical roles in maintaining soil health, nutrient cycling, and plant productivity. These fungi, which include saprotrophic decomposers, mutualistic mycorrhizal fungi, and even pathogens, form an essential part of terrestrial ecosystems. One notable example is *Glomus intraradices,* a species of arbuscular mycorrhizal fungi (AMF) from the *Glomeromycota* phylum. This fungus forms a symbiotic relationship with plant roots, enhancing nutrient and water absorption. *Glomus intraradices* extends the plant's root system through a network of hyphae that penetrate the soil, facilitating the uptake of phosphorus, nitrogen, and micronutrients. In return, the fungus receives carbohydrates from the plant. The benefits of this mutualistic association are significant in agricultural settings, as the presence of *Glomus* helps crops grow more efficiently, especially in nutrient-poor soils, while also enhancing the plant's tolerance to environmental stresses such as drought. The role of *Glomus* species in improving soil structure through aggregation and in promoting sustainable agricultural practices by reducing reliance on chemical fertilizers makes it a key player in ecosystem health.

Another important soil fungus is *Trichoderma harzianum*, an ascomycete fungus that is widely used as a biological control agent against plant pathogens. *Trichoderma* species are highly effective at outcompeting harmful fungi like *Fusarium* and *Pythium,* which are known to cause devastating root diseases in crops. By colonizing the root zone, *Trichoderma* not only protects plants from pathogenic attack but also promotes root development and overall plant vigor. Additionally, it produces enzymes and secondary metabolites that break down organic material, contributing to soil nutrient cycling and enhancing soil fertility. Beyond its agricultural applications, *Trichoderma* also improves soil health by increasing microbial diversity and activity, further supporting plant growth and resilience. Both *Glomus intraradices* and *Trichoderma harzianum* demonstrate how soil fungi contribute to plant health, soil fertility, and sustainable ecosystem management. Their roles in nutrient cycling, disease prevention, and soil structure improvement underscore the importance of maintaining a healthy and diverse fungal population in soil ecosystems.

Several species of soil fungi play crucial roles in nutrient cycling, plant health, and soil ecosystem dynamics. These fungi are essential for decomposing organic matter, forming symbiotic relationships with plants, or enhancing soil structure and fertility. Below are some key examples of important soil fungi species:

1. *Rhizopus stolonifer* (Black Bread Mold)

Rhizopus stolonifer is a member of the *Zygomycota* phylum and one of the most commonly studied soil fungi due to its role as a saprotroph. It is known for growing on decomposing plant material, particularly starchy substrates such as bread, fruits, and vegetables. In the soil, *Rhizopus* plays a significant role in breaking down organic matter and recycling nutrients, which helps maintain soil fertility. The fungus produces sporangia, structures that release spores into the environment, allowing the organism to spread rapidly in favorable conditions. *Rhizopus stolonifer* is also a model organism for studying fungal life cycles, and its enzymes have been explored for use in industrial applications, such as in the production of organic acids.

2. *Trichoderma harzianum*

Trichoderma harzianum is a widely studied soil fungus belonging to the *Ascomycota* phylum, renowned for its beneficial role in agriculture and horticulture. This fungus acts as a biological control agent, protecting plants from pathogenic fungi through mechanisms such as competition, mycoparasitism, and the production of antifungal metabolites. *Trichoderma* species are highly effective at colonizing root systems, outcompeting harmful pathogens like *Fusarium* and *Pythium*, which cause root rot and wilting diseases in plants. Additionally, *Trichoderma* enhances plant growth by stimulating root development and improving nutrient uptake, especially phosphorus and nitrogen. As a result, it is commonly used in biofertilizers and biopesticides to promote sustainable agriculture.

3. *Glomus intraradices* (Arbuscular Mycorrhizal Fungus)

Glomus intraradices belongs to the *Glomeromycota* phylum and is one of the most important species of arbuscular mycorrhizal fungi (AMF). This species forms mutualistic relationships with the roots of most terrestrial plants, particularly in agricultural systems. *Glomus intraradices* helps plants absorb phosphorus, water, and other essential nutrients from the soil by extending the root system with its extensive network of hyphae. In return, the fungus receives carbohydrates produced by the plant through photosynthesis. This symbiosis not only enhances plant growth and productivity but also improves soil structure, water retention, and resistance to environmental stresses like drought. AMF, including *Glomus intraradices,* are integral to sustainable farming and are often added to biofertilizers to reduce the reliance on chemical fertilizers.

4. *Penicillium chrysogenum*

Penicillium chrysogenum is a soil-dwelling saprotrophic fungus from the *Ascomycota* phylum, best known for its role in the production of the antibiotic penicillin. In soil ecosystems, *Penicillium* species decompose organic matter, breaking down complex compounds like cellulose and lignin, thereby contributing to the nutrient cycle. This process enriches the soil by converting organic material into simpler compounds that plants can absorb. *Penicillium chrysogenum* also plays a role in suppressing harmful soil pathogens, making it beneficial for plant health. Its ability to thrive in a variety of environmental conditions has made it a subject of extensive research in both medical and agricultural fields.

5. *Aspergillus niger*

Aspergillus niger is another saprotrophic fungus from the *Ascomycota* phylum, known for its black spores and its significant role in soil nutrient cycling. It is commonly found in soil, where it decomposes organic matter and releases nutrients like carbon, nitrogen, and phosphorus back into the ecosystem. *Aspergillus niger* produces enzymes such as amylases, cellulases, and pectinases, which are important for breaking down complex organic materials. This species is also utilized in industrial fermentation processes for

31

producing citric acid and enzymes. In agriculture, *Aspergillus niger* contributes to soil health and plant growth by decomposing organic debris and improving nutrient availability.

6. *Armillaria mellea* (Honey Fungus)

Armillaria mellea is a pathogenic fungus in the *Basidiomycota* phylum, known for causing root rot in trees and woody plants. While it acts as a pathogen, its role in forest ecosystems is complex. On one hand, it causes diseases that can kill trees, but on the other hand, it facilitates the decomposition of dead trees, contributing to nutrient cycling in forest soils. *Armillaria* can form large, long-lasting networks of underground mycelia, which can cover vast areas, making it one of the largest living organisms on Earth. The honey fungus is a critical player in forest ecology, influencing tree population dynamics, forest regeneration, and the cycling of organic matter.

7. *Fusarium oxysporum*

Fusarium oxysporum is a soil-borne fungus belonging to the *Ascomycota* phylum, known primarily for its role as a plant pathogen. This species causes wilting diseases in a variety of crops, including tomatoes, bananas, and cotton. Despite its pathogenicity, *Fusarium* also participates in decomposing plant material, contributing to organic matter breakdown in soil ecosystems. In agriculture, controlling *Fusarium* is crucial for preventing crop losses, and understanding its interactions with other soil microorganisms is key to developing sustainable agricultural practices. Some strains of *Fusarium oxysporum* are also studied for their potential use in biocontrol agents, targeting specific plant pathogens while leaving beneficial soil organisms unharmed.

8. *Beauveria bassiana*

Beauveria bassiana is a soil fungus with important applications in biological pest control. As an entomopathogenic fungus, it infects and kills a wide range of insect pests, including aphids, whiteflies, and beetles. **Beauveria bassiana** spores attach to the exoskeleton of insects, germinate, and penetrate the insect's body, ultimately killing the host. In addition to its use in pest control, this fungus contributes to the nutrient cycle by decomposing the dead insects, enriching the soil. Its role in sustainable agriculture is increasingly being recognized as a natural alternative to chemical pesticides.

9. *Mortierella species*

Mortierella species are part of the Zygomycota phylum and are widely distributed in soils, particularly in cold or acidic environments. These fungi are known for their ability to produce polyunsaturated fatty acids, such as arachidonic acid, which have commercial and health applications. In soil ecosystems, *Mortierella* species play a role in organic matter decomposition, particularly in degrading dead plant material. They are also being studied for their interactions with other soil microorganisms and their potential benefits in improving soil structure and nutrient availability.

10. *Geomyces destructans* (White-Nose Syndrome Fungus)

Geomyces destructans is a cold-loving fungus responsible for the deadly white-nose syndrome in bats. Although primarily a pathogen in bat populations, it naturally occurs in soils, particularly in caves and cold environments. *Geomyces destructans* grows on the skin of hibernating bats, causing lesions and disrupting their hibernation patterns, which leads to high mortality rates. This fungus is an example of how soil fungi can sometimes have broad ecological impacts, affecting both soil ecosystems and animal populations.

These species illustrate the critical ecological roles that soil fungi play, from nutrient cycling and plant growth to pest control and disease dynamics. The diversity of soil fungi underscores their importance in maintaining soil health, supporting plant productivity, and fostering sustainable ecosystems.

2.3.Global distribution of soil fungi and factors affecting their diversity.

The global distribution of soil fungi is highly diverse and influenced by a wide range of environmental, geographical, and biological factors. Soil fungi are found in virtually every terrestrial ecosystem, from temperate forests and grasslands to tropical rainforests, deserts, and arctic tundras. The diversity and composition of fungal communities vary significantly depending on the local climate, soil type, vegetation, and land-use practices. In tropical rainforests, for example, the warm temperatures and high moisture levels support rich fungal biodiversity, with numerous species involved in the decomposition of organic matter and the formation of symbiotic relationships with plants. In contrast, arid deserts may have lower overall fungal diversity, but the species that do thrive are highly adapted to extreme conditions, such as drought and temperature fluctuations. In colder regions like the Arctic and Antarctic, psychrophilic (cold-loving) fungi dominate, playing essential roles in organic matter decomposition in frozen soils. This vast global distribution of soil fungi is indicative of their adaptability and ecological importance across different ecosystems.

Several factors significantly affect the diversity and distribution of soil fungi. Climate is one of the most crucial drivers, as temperature, precipitation, and humidity directly influence fungal growth and activity. Warmer and wetter climates tend to support higher fungal diversity due to the availability of moisture, which is essential for spore germination and nutrient uptake. Soil pH is another critical factor, with certain fungal groups like *Ascomycota* and *Basidiomycota* favoring neutral to slightly acidic soils, while other fungi are more adapted to alkaline conditions. Land use and agricultural practices also play a major role in shaping fungal communities. Intensive farming, pesticide use, and deforestation can reduce fungal diversity by disturbing soil structure and depleting organic matter, whereas organic farming and agroforestry practices

33

that enhance soil health tend to promote fungal diversity. Finally, the presence of host plants greatly influences the distribution of mutualistic fungi, such as mycorrhizal species, which depend on plant roots for nutrients. The interactions between these factors lead to complex patterns of fungal diversity that vary across landscapes and ecosystems, reflecting both local environmental conditions and broader climatic influences.

The global distribution of soil fungi is vast and diverse, with these microorganisms found in nearly every type of terrestrial ecosystem, from dense tropical rainforests to barren deserts and the polar regions. Soil fungi play a critical role in ecosystem functioning, participating in nutrient cycling, organic matter decomposition, plant symbiosis, and even pathogen control. The distribution of soil fungi is shaped by a combination of environmental, biological, and geographical factors, resulting in distinct fungal communities across different ecosystems. In tropical rainforests, for example, warm temperatures and abundant moisture support a high diversity of fungi, particularly saprotrophic species involved in breaking down complex organic materials like lignin and cellulose. Similarly, arbuscular mycorrhizal fungi (AMF), such as **Glomus** species, are widespread in tropical and temperate regions, where they form symbiotic relationships with plant roots, facilitating nutrient uptake in exchange for carbohydrates. In contrast, desert soils harbor a more limited fungal diversity, but the species present, like xerophilic fungi, are highly specialized to withstand extreme temperatures, drought, and nutrient-poor conditions.

In arctic and alpine environments, where temperatures are low, and soils remain frozen for much of the year, psychrophilic (cold-loving) fungi dominate. These fungi are essential for the decomposition of organic matter in permafrost and tundra soils, where they play a critical role in carbon cycling. Soil fungal communities in temperate forests and grasslands tend to be dominated by species from the *Basidiomycota and Ascomycota* phyla, which contribute to the decomposition of leaf litter and woody debris. Agricultural soils, depending on management practices, can either support a diverse range of fungi, particularly mycorrhizal and saprotrophic species, or experience a decline in fungal diversity due to intensive chemical inputs and soil disturbance. The global distribution of soil fungi is not only a reflection of climate and geography but also the result of long-term evolutionary adaptations to specific environmental niches and plant communities.

Factors Affecting Soil Fungal Diversity

Several key factors influence the diversity and distribution of soil fungi globally. Climate, particularly temperature and moisture availability, is one of the most significant determinants. Fungi, being heterotrophic organisms, rely on organic matter and moisture for growth and spore dispersal, making regions with moderate to high rainfall and temperatures more conducive to fungal diversity. In tropical rainforests, the combination of heat and humidity provides ideal conditions for fungal growth and decomposition processes, while dry deserts and cold polar regions have fungi that are adapted to extreme conditions. Soil pH is another important factor; most fungi prefer neutral to slightly acidic soils, though some species can thrive in more alkaline environments. For instance, members of the Ascomycota phylum are often more abundant in acidic soils, while some *Basidiomycota* species can tolerate a broader range of pH levels.

Land use and human activities also play a significant role in shaping soil fungal communities. Agricultural practices, such as the use of fertilizers, pesticides, and tillage, can negatively impact fungal diversity by disrupting soil structure and reducing organic matter content. On the other hand, sustainable farming practices, like organic agriculture and agroforestry, tend to enhance fungal diversity by maintaining healthier soils. Vegetation and plant diversity are critical in shaping fungal populations, especially for mutualistic fungi like mycorrhizal species, which rely on plant roots for survival. In ecosystems with high plant diversity, there is typically a corresponding richness in fungal species. Soil texture and nutrient availability further influence fungal communities, as coarse, well-drained soils may favor different fungi compared to fine, water-retaining soils. Collectively, these factors interact to create complex and varied soil fungal communities across the globe, reflecting both natural environmental conditions and anthropogenic impacts.

2.4.Conclusion

Fungal biodiversity in soil represents a cornerstone of terrestrial ecosystem functioning and health, demonstrating profound implications for nutrient cycling, plant growth, and overall soil quality. The diversity of soil fungi encompasses a wide array of species, each contributing uniquely to ecological processes. From saprotrophic fungi that decompose organic matter and recycle nutrients, to mycorrhizal fungi that form symbiotic relationships with plants, and pathogenic fungi that influence plant health and ecosystem dynamics, soil fungi play multifaceted roles. Their ability to break down complex organic compounds, enhance nutrient availability, and contribute to soil structure underscores their importance in maintaining ecosystem stability and productivity. Furthermore, the intricate interactions between soil fungi and their environment highlight the necessity of preserving fungal diversity to ensure the resilience and sustainability of agricultural and natural ecosystems. As we advance in our understanding of soil fungal biodiversity, it becomes increasingly clear that effective management and conservation strategies are essential for protecting these critical organisms. Factors such as climate change, land use changes, and environmental degradation pose significant threats to fungal diversity, potentially disrupting their vital ecological functions. Integrating fungal conservation into broader soil management practices, promoting sustainable agricultural methods, and supporting research on fungal ecology and functionality are crucial steps toward safeguarding this hidden yet indispensable component of soil health. By recognizing and valuing the role of soil fungi, we can better appreciate their contributions to ecosystem services and work towards sustainable solutions that protect and enhance the rich diversity of life beneath our feet

Chapter 3

Soil Fungi and Ecosystem Functions

Soil fungi are integral to the functioning of terrestrial ecosystems, performing essential roles that influence nutrient cycling, soil structure, and plant health. As primary decomposers, soil fungi break down organic matter, including dead plant material, animal remains, and microbial biomass.

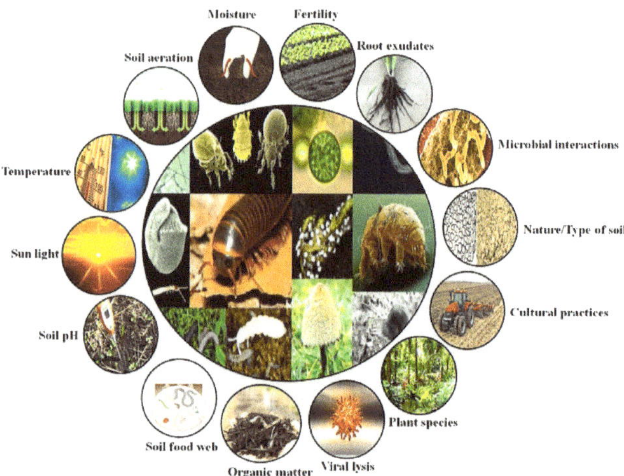

Fig 4: Biotic and Abiotic Components affecting Fungi

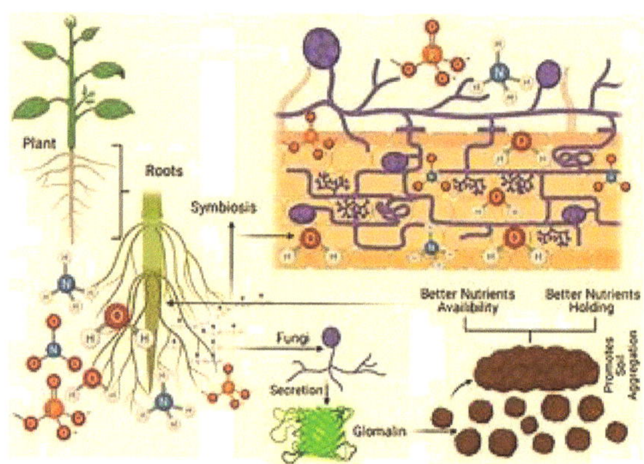

Fig 5: Role of Fungi in Ecosystem

This decomposition process releases vital nutrients such as nitrogen, phosphorus, and carbon back into the soil, making them available for uptake by plants. By converting complex organic compounds into simpler forms, fungi contribute to the formation of humus, which improves soil fertility and structure. Furthermore, fungi help in the formation and stabilization of soil aggregates, which enhances soil porosity, water infiltration, and erosion resistance. These functions are crucial for maintaining soil health and supporting plant productivity, thereby ensuring the sustainability of agricultural and natural ecosystems. The interactions between soil fungi and other soil microorganisms, such as bacteria and actinomycetes, further enhance the efficiency of nutrient cycling and organic matter decomposition.

In addition to their role in decomposition and nutrient cycling, soil fungi are vital for establishing and maintaining plant health through mutualistic relationships. Mycorrhizal fungi, including species from the genera *Glomus and Rhizophagus,* form symbiotic associations with plant roots, extending the root system through their hyphal networks. This mycorrhizal network significantly enhances the plant's ability to absorb essential nutrients, particularly phosphorus and nitrogen, and improves its tolerance to environmental stresses such as drought and soil salinity. In return, the plant provides the fungi with carbohydrates produced through photosynthesis. This mutualistic relationship not only benefits individual plants but also has broader implications for ecosystem functioning, as it influences plant community dynamics and productivity. Additionally, certain soil fungi play a role in suppressing soil-borne pathogens through competitive exclusion and the production of antifungal compounds, further contributing to plant health and resilience. By performing these diverse functions, soil fungi underpin the stability and productivity of ecosystems, highlighting the importance of preserving their diversity and ecological roles.

Soil fungi play pivotal roles in ecosystem functioning, significantly influencing nutrient cycling, soil structure, and plant health. Their contributions are essential for maintaining soil fertility and supporting plant productivity, thereby ensuring the stability and sustainability of terrestrial ecosystems.

Nutrient Cycling and Decomposition

One of the primary functions of soil fungi is their role in the decomposition of organic matter. Soil fungi, including saprotrophic species such as those from the genera *Penicillium, Aspergillus,* and *Fusarium,* break down complex organic materials like leaf litter, wood, and dead organisms. This process is crucial for the release of essential nutrients—such as nitrogen, phosphorus, and carbon—into the soil. Fungi decompose cellulose, lignin, and other recalcitrant compounds, converting them into simpler forms that plants can readily absorb. By transforming organic matter into humus, fungi contribute to soil structure, increasing its ability to retain water and nutrients. The formation of humus also enhances soil aggregation, which improves soil aeration, reduces erosion, and enhances root penetration

The breakdown of organic material by soil fungi is a key component of the carbon cycle. Fungi release carbon dioxide into the atmosphere as a byproduct of decomposition, which is a natural part of the global carbon cycle. Additionally, fungi help to sequester carbon in soil organic matter, which can mitigate the effects of climate change by storing carbon in the soil. This function is particularly important in forest and agricultural soils, where organic matter accumulation and breakdown play significant roles in soil health and carbon dynamics.

Mycorrhizal Associations and Plant Health

In addition to decomposition, soil fungi form crucial mutualistic relationships with plant roots through mycorrhizal associations. Mycorrhizal fungi, including those from the genera **_Glomus, Rhizophagus_**, and **_Cenococcum_**, extend the plant's root system through an extensive network of hyphae. This fungal network significantly enhances the plant's ability to absorb nutrients, particularly phosphorus, which is often limiting in many soils. The mycorrhizal hyphae increase the surface area for nutrient uptake and can access nutrients that are otherwise inaccessible to plant roots. In exchange, the fungi receive carbohydrates and other organic compounds from the plant.

These mycorrhizal associations also improve plant resistance to environmental stresses such as drought, soil salinity, and heavy metal toxicity. By enhancing nutrient uptake and improving soil structure, mycorrhizal fungi help plants to better withstand challenging conditions and increase their overall resilience. Moreover, mycorrhizal fungi contribute to the formation of soil aggregates and the stabilization of soil structure, which further supports plant growth and reduces soil erosion. This symbiotic relationship is vital not only for the health of individual plants but also for the functioning of entire ecosystems, influencing plant community composition, productivity, and biodiversity.

Pathogen Suppression and Ecosystem Stability

Soil fungi also play a role in suppressing soil-borne pathogens, contributing to plant health and ecosystem stability. Some fungi produce antibiotics and other bioactive compounds that inhibit the growth of harmful pathogens. For example, certain strains of **_Trichoderma_** and **_Beauveria_** are used in biological control to manage plant diseases caused by fungi and insects, respectively. By competing with pathogenic fungi for resources and space, or by directly parasitizing them, beneficial soil fungi help to maintain a balance in the microbial community and reduce disease incidence.

Soil fungi are integral to ecosystem functions through their roles in nutrient cycling, soil structure formation, plant health, and pathogen suppression. Their diverse functions underscore their importance in maintaining soil fertility, supporting plant productivity, and ensuring ecosystem sustainability. Understanding and preserving soil fungal diversity is crucial for the health and resilience of terrestrial ecosystems, highlighting the need for continued research and conservation efforts.

3.1.Role of fungi in nutrient cycling (decomposition, nitrogen fixation, phosphorus solubilization).

Fungi play a critical role in nutrient cycling within terrestrial ecosystems by participating in the decomposition of organic matter, which is fundamental for the release and recycling of essential nutrients. As decomposers, fungi break down complex organic materials such as leaf litter, wood, and dead organisms, transforming them into simpler forms that can be assimilated by plants. This decomposition process involves various types of fungi, including saprotrophic species like *Aspergillus* and *Penicillium*, which secrete extracellular enzymes to degrade cellulose, lignin, and other polymers. By breaking down these complex compounds, fungi release nutrients such as nitrogen, phosphorus, and potassium back into the soil. These nutrients, once bound in organic matter, become available in forms that plants can absorb through their roots. This process not only recycles essential nutrients but also contributes to soil formation and structure, as decomposed organic matter forms humus, which enhances soil fertility and its ability to retain water and nutrients. Additionally, the breakdown of organic matter by fungi influences the global carbon cycle, as carbon is released into the atmosphere in the form of carbon dioxide, and a portion is sequestered in the soil as stable organic matter.

Fungi also play a significant role in the nitrogen cycle, particularly through their involvement in processes such as nitrogen mineralization and mycorrhizal symbioses. In nitrogen mineralization, fungi convert organic nitrogen from decaying plant and animal matter into inorganic forms, such as ammonium and nitrate, which plants can readily uptake. This process is essential for maintaining soil nitrogen levels and supporting plant growth. Mycorrhizal fungi, such as those from the genera *Glomus and Rhizophagus*, further contribute to nutrient cycling by forming symbiotic relationships with plant roots. These fungi extend the plant's root system through a network of hyphae that increases nutrient absorption efficiency, particularly for phosphorus, which is often limited in soils. In exchange, the fungi receive carbohydrates and other organic compounds from the plant. This mutualistic relationship enhances the plant's nutrient uptake and overall health, influencing plant community dynamics and ecosystem productivity. By facilitating the movement of nutrients between the soil and plants, fungi are crucial for the efficiency and sustainability of nutrient cycling in terrestrial ecosystems.

Fungi are central to nutrient cycling in ecosystems, primarily through their role in decomposition. Decomposition is the process by which organic matter, such as plant litter, dead animals, and microbial biomass, is broken down into simpler compounds. This process is vital for recycling nutrients and maintaining soil fertility.

Decomposition Process

Fungi, particularly saprotrophic fungi, are essential decomposers in terrestrial ecosystems. They utilize extracellular enzymes to break down complex organic compounds such as cellulose, lignin, and chitin found in plant and animal tissues. This enzymatic degradation involves several key stages:

1. **Enzyme Production**: Decomposing fungi produce a range of enzymes that catalyze the breakdown of complex organic molecules. Cellulases degrade cellulose, which is a major component of plant cell walls. Ligninases break down lignin, another key structural component of plant tissues. Chitinases target chitin, a component of fungal cell walls and exoskeletons of arthropods.

2. **Substrate Colonization**: Fungal hyphae penetrate the organic matter, expanding through the substrate and increasing the surface area for enzymatic activity. This colonization is crucial for efficient decomposition, as it allows fungi to access and decompose otherwise inaccessible nutrients.

3. **Nutrient Release**: As fungi decompose organic materials, they release nutrients such as nitrogen, phosphorus, and potassium back into the soil. Nitrogen is often released in the form of ammonium (NH_4^+) or nitrate (NO_3^-), while phosphorus is released as orthophosphate (PO_4^{3-}). These nutrients are then available for uptake by plants, contributing to soil fertility and plant growth.

4. **Formation of Humus**: The decomposition process also contributes to the formation of humus, a stable organic component of soil. Humus is rich in nutrients and improves soil structure by enhancing soil aggregation, water retention, and aeration. This stable organic matter serves as a reservoir of nutrients, gradually releasing them to plants over time.

Ecological Significance

The role of fungi in decomposition has significant ecological implications:

- **Nutrient Cycling**: By breaking down organic matter, fungi play a crucial role in nutrient cycling. The release of nutrients from decomposed organic material ensures that essential elements are continually recycled within the ecosystem, supporting plant growth and productivity.

- **Soil Fertility**: The formation of humus and the release of nutrients contribute to soil fertility. Healthy, fertile soils support diverse plant communities and enhance ecosystem resilience.

- **Carbon Cycle**: Fungal decomposition is a key component of the global carbon cycle. As organic matter is decomposed, carbon is released into the atmosphere as carbon dioxide (CO_2) through microbial respiration. However, a portion of the carbon is sequestered in soil organic matter, contributing to long-term carbon storage and influencing climate change dynamics.

- **Ecosystem Dynamics**: Decomposition influences plant community dynamics by affecting nutrient availability. The diversity and activity of decomposing fungi can impact the types of plants that thrive in a given area, thereby shaping plant community composition and ecosystem structure.

Fungi are indispensable to nutrient cycling through their role in decomposition. Their ability to break down complex organic materials, release essential nutrients, and contribute to soil formation highlights their importance in maintaining soil health, fertility, and overall ecosystem functioning.

Nitrogen Fixation by Fungi

Nitrogen fixation is a critical process in nutrient cycling that converts atmospheric nitrogen (N_2) into forms usable by plants. While nitrogen fixation is predominantly associated with certain bacteria, fungi, particularly those involved in mycorrhizal associations, can play a supporting role in nitrogen cycling.

1. **Mycorrhizal Associations**: Mycorrhizal fungi, such as those from the genus **Glomus**, form symbiotic relationships with plant roots. While mycorrhizal fungi themselves do not fix atmospheric nitrogen, they can interact with nitrogen-fixing bacteria within the rhizosphere. For example, in some ecosystems, mycorrhizal fungi can support the growth of actinobacteria, which are capable of nitrogen fixation. This indirect role enhances the availability of nitrogen for plants.

2. **Decomposers and Nitrogen Mineralization**: Fungi also contribute to nitrogen cycling through the decomposition of organic matter. As saprotrophic fungi break down complex organic compounds, they convert organic nitrogen from dead plant and animal material into inorganic forms, such as ammonium (NH_4^+). This process, known as nitrogen mineralization, is crucial for making nitrogen available to plants. Fungi release ammonium through their enzymatic activities, and this ammonium can be further converted into nitrate (NO_3^-) by nitrifying bacteria in the soil.

Phosphorus Solubilization by Fungi

Phosphorus is a vital nutrient for plant growth, but it is often present in soils in forms that are not readily available to plants. Fungi, particularly mycorrhizal fungi, play a key role in making phosphorus more accessible through their solubilization activities.

1. **Mycorrhizal Fungi and Phosphorus Uptake**: Mycorrhizal fungi, including arbuscular mycorrhizal fungi (AMF) such as **Glomus** and **Rhizophagus**, form symbiotic relationships with plant roots. The fungal hyphae extend into the soil, creating a vast network that increases the surface area for nutrient absorption. One of the critical functions of these fungi is to solubilize inorganic phosphorus from soil minerals. They secrete organic acids, such as citric acid and oxalic acid, which acidify the soil microenvironment and dissolve phosphorus from insoluble mineral forms like apatite. This process releases orthophosphate (PO_4^{3-}), making it available for plant uptake.

2. **Direct and Indirect Mechanisms**: In addition to organic acid secretion, mycorrhizal fungi can also employ other mechanisms to solubilize phosphorus. They produce enzymes, such as phosphatases, that break down organic phosphorus compounds into inorganic forms that plants can absorb. Furthermore, the mycorrhizal hyphae can explore larger soil volumes compared to plant roots alone,

accessing phosphorus that is otherwise out of reach for plants. By enhancing phosphorus availability, mycorrhizal fungi support plant growth and productivity, particularly in phosphorus-poor soils.

Ecological Significance

- **Enhanced Plant Nutrition**: Both nitrogen and phosphorus are essential for plant growth and development. The role of fungi in nitrogen mineralization and phosphorus solubilization directly impacts plant health, productivity, and ecosystem dynamics. By facilitating nutrient uptake, fungi help plants thrive in various environments, from nutrient-poor soils to highly productive agricultural systems.

- **Soil Health and Fertility**: The activities of fungi in nitrogen cycling and phosphorus solubilization contribute to overall soil health and fertility. By ensuring the availability of these critical nutrients, fungi support soil structure and fertility, which in turn enhances soil ecosystem functions and resilience.

- **Ecosystem Productivity**: By improving nutrient availability, fungi play a role in influencing plant community composition and ecosystem productivity. Healthy, nutrient-rich soils support diverse plant communities, which can affect ecosystem stability, biodiversity, and productivity.

Fungi contribute to nutrient cycling through both nitrogen mineralization and phosphorus solubilization. Their roles in making essential nutrients available to plants are crucial for maintaining soil health, supporting plant growth, and ensuring the sustainability of ecosystems.

Phosphorus solubilization by fungi

Phosphorus solubilization by fungi is a crucial aspect of nutrient cycling in terrestrial ecosystems. Phosphorus is a key nutrient for plant growth, but it is often present in soils in forms that are not readily available to plants. Soil fungi, particularly mycorrhizal fungi, play a significant role in transforming phosphorus into forms that can be absorbed by plant roots, thus enhancing soil fertility and plant productivity.

Mechanisms of Phosphorus Solubilization

1. **Organic Acid Production**: Mycorrhizal fungi, including arbuscular mycorrhizal fungi (AMF) such as those from the genus **Glomus** and **Rhizophagus**, secrete organic acids into the soil. These acids, such as citric acid, oxalic acid, and malic acid, lower the pH of the soil microenvironment around the fungal hyphae. The acidification of the soil dissolves phosphorus from mineral forms such as apatite (calcium phosphate), which is otherwise insoluble. By converting insoluble phosphorus into orthophosphate (PO_4^{3-}), these fungi make the nutrient available for plant uptake.

2. **Enzyme Production**: Mycorrhizal fungi also produce extracellular enzymes, such as phosphatases, that help to release phosphorus from organic compounds in the soil. Phosphatases break down organic phosphorus compounds, such as phytate (a common form of phosphorus in plant residues), into inorganic forms that can be absorbed by plants. This enzymatic activity is particularly important in soils where phosphorus is bound in organic matter.

3. **Hyphal Exploration**: The extensive network of fungal hyphae extends far beyond the reach of plant roots, allowing fungi to access phosphorus that is otherwise out of reach. This mycelial network increases the effective root surface area for nutrient absorption and enables the uptake of phosphorus from deeper and less accessible soil layers.

4. **Formation of Mycorrhizal Networks**: In addition to their direct effects on phosphorus availability, mycorrhizal fungi can form networks that connect multiple plants. This mycorrhizal network facilitates nutrient exchange between plants, allowing for the transfer of phosphorus from one plant to another. This is particularly beneficial in ecosystems where phosphorus is scarce, as it enables plants to share resources and improve overall nutrient utilization.

Ecological and Agricultural Implications

- **Enhanced Plant Growth**: By solubilizing phosphorus and making it available to plants, fungi play a crucial role in supporting plant growth and development. Phosphorus is essential for energy transfer, photosynthesis, and nucleic acid synthesis. Adequate phosphorus availability enhances plant vigor, root development, and overall productivity.

- **Soil Fertility**: The ability of fungi to solubilize phosphorus contributes to soil fertility. Increased phosphorus availability supports higher plant yields and improves the nutritional quality of crops. This is particularly important in phosphorus-poor soils, which are common in many agricultural and natural ecosystems.

- **Sustainable Agriculture**: Mycorrhizal fungi can reduce the need for synthetic phosphorus fertilizers, which are often costly and can lead to environmental issues such as eutrophication of water bodies. By promoting natural phosphorus cycling, fungi support sustainable agricultural practices and reduce the environmental impact of farming.

- **Ecosystem Functioning**: In natural ecosystems, the role of fungi in phosphorus solubilization helps to maintain soil health and ecosystem stability. By ensuring that plants have access to essential nutrients, fungi support diverse plant communities and contribute to ecosystem resilience.

Fungi play a vital role in phosphorus solubilization through mechanisms such as organic acid production, enzyme activity, and hyphal exploration. Their contributions to nutrient cycling enhance soil fertility, support plant growth, and promote sustainable agricultural and ecological practices. Understanding and leveraging the role of fungi in phosphorus cycling is crucial for improving soil health and ensuring the sustainability of terrestrial ecosystems.

3.2.Symbiotic relationships between fungi and plants (mycorrhizae).

Symbiotic relationships between fungi and plants are fundamental to many terrestrial ecosystems and involve a range of interactions that significantly benefit both partners. The most well-known of these relationships is the mycorrhizal association, where fungi form mutualistic partnerships with plant roots. In this symbiosis, mycorrhizal fungi extend their hyphae into the soil, creating an expansive network that increases the surface area for nutrient absorption. This network allows the fungi to access soil resources, particularly phosphorus, that are otherwise difficult for plants to obtain. The fungi solubilize phosphorus from soil minerals and organic matter, converting it into a form that is readily absorbed by the plant. In return, the plant provides the fungi with carbohydrates and other organic compounds produced through photosynthesis. This exchange supports the fungi's growth and reproductive processes while enhancing the plant's nutrient uptake and overall health. This mutualistic relationship not only improves plant nutrition but also enhances soil structure, water retention, and resistance to soil-borne pathogens.

In addition to mycorrhizal associations, another significant symbiotic relationship between fungi and plants is the endophytic symbiosis. Endophytic fungi live within plant tissues without causing harm, often residing in roots, stems, or leaves. These fungi can enhance plant health by producing secondary metabolites that act as antimicrobial agents, protecting the plant from pathogens and herbivores. Some endophytes also contribute to nutrient acquisition by assisting in the breakdown of organic matter within the plant's environment, making nutrients more available. Additionally, endophytes can improve plant tolerance to environmental stresses such as drought, salinity, and heavy metal toxicity. By promoting plant growth and resilience through these mechanisms, endophytic fungi play a crucial role in the health and productivity of their host plants. Both mycorrhizal and endophytic symbioses underscore the importance of fungal-plant interactions in supporting ecosystem functions and sustaining plant communities.

Mycorrhizae represent one of the most crucial and widespread symbiotic relationships between fungi and plants. These associations are essential for nutrient uptake, soil health, and overall ecosystem stability. Mycorrhizae can be broadly categorized into several types based on the nature of their interaction with plant roots, including arbuscular mycorrhizae (AM), ectomycorrhizae (ECM), and ectendomycorrhizae. Each type of mycorrhizae has distinct features and functions, but all share the common benefit of enhancing nutrient acquisition for plants.

Arbuscular Mycorrhizae (AM)

Arbuscular mycorrhizae, also known as endomycorrhizae, are characterized by the formation of structures called arbuscules and vesicles within plant root cells. These fungi belong to the phylum **Glomeromycota** and form symbiotic relationships with approximately 80-90% of terrestrial plants, including many crops and native species. In AM, the fungal hyphae penetrate the plant root cells, forming arbuscules, which are tree-like structures where nutrient exchange occurs. These structures facilitate the transfer of nutrients, particularly phosphorus, from the soil to the plant. Phosphorus is essential for various plant processes, including energy transfer and nucleic acid synthesis. The fungi also form vesicles, which act as storage organs for lipids and nutrients. In exchange for these nutrients, the plant provides the fungi with carbohydrates and other organic compounds derived from photosynthesis. AM fungi extend their hyphae into the soil, greatly increasing the effective root surface area and enhancing the plant's ability to access nutrients and water. This association not only improves nutrient uptake but also enhances plant resistance to environmental stresses such as drought and soil salinity.

Ectomycorrhizae (ECM)

Ectomycorrhizae are another major type of mycorrhizal association, primarily involving fungi from the phyla **Basidiomycota** and **Ascomycota**. Unlike AM fungi, ECM fungi do not penetrate the plant root cells but instead form a dense network of hyphae around the outside of root cells, creating a sheath or mantle. This fungal mantle extends into the soil, enhancing the plant's ability to absorb water and nutrients, particularly nitrogen and phosphorus. ECM fungi also form a structure called the Hartig net, a network of fungal hyphae that penetrates the root cell walls but does not enter the cells themselves. This network facilitates nutrient exchange between the plant and the soil. ECM associations are particularly common in temperate and boreal forests and are essential for the health of many tree species, such as pines, oaks, and birches. The symbiotic relationship benefits the plant by improving nutrient uptake, particularly in nutrient-poor soils, and enhances its ability to tolerate soil pathogens and environmental stresses. In return, the plant supplies the ECM fungi with carbohydrates and other organic compounds.

Ecological and Agricultural Role

Both arbuscular and ectomycorrhizal fungi play critical roles in ecosystem functioning and agriculture. By enhancing nutrient availability and improving soil structure, mycorrhizal fungi support plant growth, productivity, and resilience. They also contribute to soil health by promoting the formation of soil aggregates, which improves water infiltration and reduces erosion. In agriculture, mycorrhizal inoculants are used to enhance crop yields and reduce the need for synthetic fertilizers, thereby promoting sustainable farming practices. Additionally, mycorrhizal fungi can help plants recover from disturbances and environmental stressors, contributing to ecosystem stability and resilience.

Mycorrhizae are a vital component of plant-fungal interactions, significantly influencing nutrient uptake, soil health, and ecosystem stability. Whether through arbuscular mycorrhizae's internal symbiosis or ectomycorrhizae's external association, these fungal partnerships provide essential benefits to plants and contribute to the overall functioning of terrestrial ecosystems.

Mycorrhizae, the symbiotic associations between fungi and plant roots, are of paramount importance in various ecological, agricultural, and environmental contexts. These relationships enhance plant growth, soil health, and ecosystem stability in numerous ways. Here's a detailed exploration of their significance:

Enhancing Nutrient Uptake

1. Phosphorus Acquisition: One of the most critical roles of mycorrhizae is the enhancement of phosphorus uptake. Phosphorus is often present in soil in forms that are not readily available to plants. Mycorrhizal fungi, particularly arbuscular mycorrhizal fungi (AMF) and ectomycorrhizal fungi (ECMF), extend their hyphal networks into the soil, increasing the effective root surface area. They secrete organic acids and enzymes that solubilize phosphorus from insoluble minerals and organic matter, making it available to plants. This is crucial for plant processes such as energy transfer, photosynthesis, and nucleic acid synthesis. Phosphorus acquisition by mycorrhizal fungi is a critical process that significantly enhances plant nutrition and ecosystem productivity. Phosphorus is a vital nutrient for plants, involved in energy transfer, photosynthesis, and nucleic acid synthesis. However, phosphorus is often present in soils in forms that are not readily accessible to plants. Mycorrhizal fungi play a pivotal role in converting phosphorus into a form that plants can absorb. Here's a detailed examination of how mycorrhizal fungi enhance phosphorus acquisition:

Mechanisms of Phosphorus Solubilization

1. **Hyphal Exploration and Extension**:

     Increased Surface Area: Mycorrhizal fungi, particularly arbuscular mycorrhizal fungi (AMF) and ectomycorrhizal fungi (ECMF), extend their hyphal networks into the soil, significantly increasing the effective surface area for nutrient absorption. These fungal hyphae can explore a larger volume of soil compared to plant roots alone, reaching phosphorus that is not accessible to roots.

     Hyphal Penetration: In arbuscular mycorrhizae, the fungal hyphae penetrate the plant root cells to form arbuscules, which are specialized structures where nutrient exchange occurs. This intimate contact with plant roots allows fungi to directly interface with the soil environment and access phosphorus that is otherwise out of reach.

2. **Organic Acid Secretion**:

Acidification of the Soil: Mycorrhizal fungi secrete organic acids such as citric acid, oxalic acid, and malic acid into the soil. These acids lower the pH around the fungal hyphae, which helps dissolve phosphorus from insoluble mineral forms like apatite (calcium phosphate).

Phosphorus Solubilization: The organic acids dissolve insoluble phosphorus compounds, converting them into orthophosphate (PO_4^{3-}), which is the form of phosphorus that plants can readily absorb. This process effectively increases the availability of phosphorus in the soil, making it more accessible to plants.

3. **Enzymatic Activity**:

    Phosphatase Production: Mycorrhizal fungi produce extracellular enzymes, such as phosphatases, which break down organic phosphorus compounds. These enzymes hydrolyze organic forms of phosphorus, such as phytate (a common form of phosphorus in plant residues), releasing inorganic phosphate into the soil solution.

    Enhanced Nutrient Availability: By converting organic phosphorus into inorganic forms, these enzymes help to make phosphorus available for plant uptake. This process is particularly important in soils rich in organic matter where phosphorus is bound in organic compounds.

Impact on Plant Nutrition and Soil Health

1. **Improved Plant Growth**:

    Enhanced Phosphorus Uptake: By increasing the availability of phosphorus, mycorrhizal fungi significantly enhance plant growth and development. Phosphorus is crucial for energy transfer, photosynthesis, and the formation of nucleic acids, which are essential for cell division and plant metabolism.

    Increased Root Development: The presence of mycorrhizal fungi promotes healthier and more extensive root systems, further improving the plant's ability to access nutrients and water from the soil.

2. **Soil Structure and Fertility**:

    Formation of Soil Aggregates: Mycorrhizal fungi contribute to the formation of soil aggregates by secreting polysaccharides that bind soil particles together. This improves soil structure, water infiltration, and aeration, which enhances overall soil health.

    Nutrient Cycling: By breaking down organic matter and solubilizing phosphorus, mycorrhizal fungi play a role in nutrient cycling and soil fertility. Their activities help maintain the nutrient balance in the soil, supporting sustainable plant growth.

Ecological and Agricultural Significance

1. **Ecosystem Functioning**:

   Support for Plant Diversity: Mycorrhizal fungi contribute to plant biodiversity by improving nutrient availability in various soil types. This supports a diverse range of plant species and contributes to ecosystem stability.

   Resilience to Stress: Mycorrhizal associations can enhance plant resilience to environmental stresses such as drought and soil salinity, which is crucial for maintaining ecosystem health under changing conditions.

2. **Agricultural Applications**:

   Reduced Fertilizer Use: Mycorrhizal inoculants are used in agriculture to enhance crop yields and reduce the reliance on synthetic phosphorus fertilizers. This promotes more sustainable farming practices and reduces environmental impacts such as eutrophication.

   Restoration Projects: In ecological restoration, mycorrhizal fungi are employed to improve soil fertility and support the establishment of vegetation in degraded or disturbed soils.

Mycorrhizal fungi play a critical role in phosphorus acquisition by plants through mechanisms such as hyphal extension, organic acid secretion, and enzymatic activity. Their contributions enhance plant growth, improve soil structure, and support ecosystem health, making them indispensable partners in both natural and agricultural systems.

2. Other Nutrient Uptake: In addition to phosphorus, mycorrhizal fungi also enhance the uptake of other essential nutrients, including nitrogen, potassium, calcium, and micronutrients. For instance, ECMF are particularly effective at accessing nitrogen, which is crucial for protein synthesis and plant growth. Mycorrhizal networks can access deeper soil layers and provide nutrients that might otherwise be unavailable to plants. Mycorrhizal fungi significantly enhance the uptake of several key nutrients besides phosphorus, including nitrogen, potassium, calcium, and micronutrients. These fungi form symbiotic associations with plant roots, extending their hyphal networks into the soil and increasing nutrient availability. Here's a detailed examination of how mycorrhizal fungi contribute to the uptake of various essential nutrients:

Nitrogen Uptake

1. **Nutrient Acquisition**:

   Hyphal Network Expansion: Mycorrhizal fungi, particularly ectomycorrhizal fungi (ECMF) and some arbuscular mycorrhizal fungi (AMF), extend their hyphal networks into the soil to access nitrogen sources. These fungi can absorb nitrogen compounds directly from the soil solution and transfer them to the plant. This is particularly valuable in soils where nitrogen is in limited or less accessible forms.

Symbiotic Benefits: Some mycorrhizal fungi can assimilate nitrogen in the form of ammonium (NH_4^+) or nitrate (NO_3^-) from the soil and then provide it to the plant in a readily usable form. This contributes to plant growth by supporting protein synthesis and other metabolic processes.

2. **Interactions with Soil Microorganisms**:

Nitrogen Fixation: While mycorrhizal fungi themselves do not fix atmospheric nitrogen, they often interact with other soil microorganisms, such as nitrogen-fixing bacteria. These interactions can indirectly benefit plants by enhancing the overall nitrogen availability in the soil. Mycorrhizal fungi may facilitate the uptake of nitrogen fixed by these bacteria, making it available to plants.

Potassium Uptake

1. **Root-Soil Interface**:

Enhanced Absorption: Mycorrhizal fungi improve potassium uptake by increasing the surface area of the root-soil interface. The fungal hyphae explore a larger volume of soil, accessing potassium ions (K^+) that may be less accessible to plant roots alone.

K^+ Mobilization: Mycorrhizal fungi can also affect the mobilization and availability of potassium in the soil. They release organic acids and other compounds that can help dissolve potassium-containing minerals, making potassium more available to plants.

2. **Improved Plant Health**:

Stress Tolerance: Adequate potassium availability is crucial for maintaining cell turgor, enzyme activation, and photosynthesis. Mycorrhizal fungi enhance potassium uptake, contributing to improved plant health and stress tolerance, particularly under adverse conditions such as drought.

Calcium Uptake

1. **Soil Calcium Availability**:

Calcium Acquisition: Mycorrhizal fungi, especially AMF, enhance calcium uptake by extending their hyphal networks into the soil. Calcium ions (Ca^{2+}) are often bound in soil minerals or organic matter, and mycorrhizal fungi help to make these ions more accessible to plant roots.

Soil Interaction: The interaction between fungal hyphae and soil particles can release calcium from insoluble forms, improving its availability to plants. This is particularly important in soils with low calcium content or where calcium is tightly bound.

2. **Impact on Plant Health**:

Cell Wall Integrity: Calcium is essential for maintaining cell wall integrity and stability. Mycorrhizal fungi enhance calcium uptake, supporting the structural health of plant tissues and contributing to overall plant vigor.

Micronutrient Uptake

1. **Trace Elements**:

Enhanced Micronutrient Access: Mycorrhizal fungi improve the uptake of essential micronutrients such as iron (Fe), zinc (Zn), copper (Cu), and manganese (Mn). These elements are crucial for various enzymatic and physiological functions within the plant.

Mechanisms of Solubilization: Mycorrhizal fungi release organic acids and enzymes that can solubilize micronutrients from soil particles and organic matter, making them more available to plants.

2. **Nutrient Balance**:

Micronutrient Interactions: The enhanced uptake of micronutrients by mycorrhizal fungi helps maintain a balanced nutrient profile within the plant, preventing deficiencies and supporting optimal growth and development.

Ecological and Agricultural Significance

1. **Soil Health and Fertility**:

Nutrient Cycling: Mycorrhizal fungi contribute to nutrient cycling by breaking down organic matter and mobilizing nutrients. Their activities improve soil fertility and support healthy plant communities.

Soil Structure: By forming mycorrhizal networks, fungi contribute to soil aggregation and structure, which enhances water retention and reduces erosion.

2. **Sustainable Agriculture**:

Reduced Fertilizer Dependency: Mycorrhizal inoculants can reduce the need for synthetic fertilizers by improving nutrient uptake. This supports sustainable farming practices and reduces environmental impacts associated with fertilizer use.

Crop Yields and Health: Enhanced nutrient uptake leads to increased crop yields and improved plant health. Mycorrhizal fungi can be used to support the growth of crops in nutrient-poor or degraded soils.

Mycorrhizal fungi play a vital role in enhancing the uptake of various essential nutrients, including nitrogen, potassium, calcium, and micronutrients. Their ability to extend hyphal networks, release organic acids, and interact with soil particles contributes to improved nutrient availability, plant health, and soil fertility. These benefits are crucial for maintaining ecosystem health and supporting sustainable agricultural practices.

Improving Soil Health

1. Soil Structure: Mycorrhizal fungi contribute to soil structure by promoting the formation of soil aggregates. The fungal hyphae excrete polysaccharides that bind soil particles together, creating a more stable soil structure. This improves soil aeration, water infiltration, and reduces erosion, leading to healthier soil ecosystems. Soil structure, the arrangement of soil particles into aggregates or clumps, is a critical component of soil health and fertility. Mycorrhizal fungi play a pivotal role in enhancing soil structure through several mechanisms. Their influence on soil aggregation, water infiltration, aeration, and erosion control contributes significantly to maintaining and improving soil health.

Formation of Soil Aggregates

1. **Fungal Hyphal Networks**:

    Hyphal Binding: Mycorrhizal fungi form extensive networks of hyphae that extend into the soil and interact with soil particles. These fungal hyphae secrete polysaccharides and other binding agents that help to bind soil particles together into aggregates. This process enhances the formation and stability of soil aggregates, which are crucial for maintaining soil structure.

    Organic Matter Incorporation: Mycorrhizal fungi contribute to the incorporation of organic matter into the soil. As they break down organic materials, such as plant residues and leaf litter, they produce organic compounds that help to stabilize soil aggregates. This organic matter also acts as a glue, binding soil particles together and forming stable aggregates.

2. **Improved Soil Porosity**:

    Increased Air Space: The formation of soil aggregates increases soil porosity, which improves the air space within the soil. This enhanced porosity facilitates better aeration, allowing for more efficient gas exchange between the soil and the atmosphere. Adequate soil aeration is essential for the health of plant roots and soil microorganisms.

Water Infiltration and Retention: Improved soil structure enhances water infiltration and reduces surface runoff. Aggregates create larger pore spaces in the soil, which allows water to penetrate more deeply and reduces the risk of erosion. Additionally, well-aggregated soils have better water-holding capacity, which helps to retain moisture for plant use.

Soil Erosion Control

1. **Erosion Prevention**:

   Stabilization of Soil Surface: The presence of mycorrhizal fungi and their hyphal networks helps to stabilize the soil surface, reducing the risk of erosion. The fungal hyphae intertwine with soil particles and organic matter, creating a more stable surface that resists the forces of wind and water erosion.

   Root Support: Mycorrhizal fungi improve the health and development of plant roots, which further contributes to soil stabilization. Healthy plant roots, supported by mycorrhizal associations, anchor the soil in place, reducing the potential for erosion.

2. **Erosion Mitigation**:

   Vegetative Cover: Mycorrhizal fungi enhance plant growth and establishment, leading to more robust vegetative cover. This cover helps to protect the soil surface from erosion by reducing the impact of rainfall and wind. The extensive root systems of plants supported by mycorrhizal fungi also contribute to soil stability.

Impact on Soil Fertility

1. **Nutrient Cycling**:

   Enhanced Decomposition: Mycorrhizal fungi contribute to the decomposition of organic matter in the soil. This process releases nutrients, such as nitrogen, phosphorus, and potassium, back into the soil, improving soil fertility. The breakdown of organic matter also contributes to the formation of stable soil aggregates.

   Nutrient Availability: By improving soil structure and nutrient cycling, mycorrhizal fungi enhance the availability of essential nutrients for plant uptake. This results in better plant growth and productivity.

2. **Soil Organic Matter**:

   Increased Organic Matter Content: The activity of mycorrhizal fungi increases the organic matter content of the soil. Organic matter is a key component of soil structure, as it helps to bind soil particles together and improve soil fertility. The presence of organic matter also enhances soil moisture retention and provides a habitat for beneficial soil microorganisms.

Ecological and Agricultural Significance

1. **Ecosystem Health**:

   Support for Plant Communities: Improved soil structure supports the health and diversity of plant communities. By enhancing soil aeration, water infiltration, and nutrient availability, mycorrhizal fungi contribute to the establishment and growth of a wide range of plant species.

   Soil Microbial Activity: A well-structured soil environment supports a diverse and active microbial community. Mycorrhizal fungi contribute to the overall health of the soil ecosystem by providing habitats for beneficial microorganisms and enhancing nutrient cycling.

2. **Sustainable Agriculture**:

   Reduced Soil Compaction: Improved soil structure reduces soil compaction, which can hinder root growth and water infiltration. By promoting soil aggregation and enhancing porosity, mycorrhizal fungi help to create a more favorable environment for plant growth.

   Resilience to Environmental Stress: Well-structured soils are more resilient to environmental stresses such as drought and erosion. Mycorrhizal fungi contribute to soil health and stability, supporting sustainable agricultural practices and reducing the need for external inputs.

Mycorrhizal fungi play a crucial role in improving soil health by enhancing soil structure. Through the formation of soil aggregates, increased porosity, erosion control, and nutrient cycling, they contribute to better soil fertility, water management, and overall ecosystem health. Their impact on soil structure supports both natural ecosystems and agricultural systems, highlighting their importance in maintaining and restoring soil health.

2. Organic Matter Decomposition: Some mycorrhizal fungi participate in the decomposition of organic matter, contributing to nutrient cycling. They break down organic materials, releasing nutrients back into the soil, which supports plant growth and maintains soil fertility. Organic matter decomposition is a vital process in soil health, significantly influencing soil fertility, structure, and overall ecosystem functioning. Mycorrhizal fungi play an essential role in this process by breaking down organic materials and facilitating nutrient cycling. Their activities help transform organic matter into forms that are beneficial for plants and contribute to the sustainability of soil ecosystems.

Role of Mycorrhizal Fungi in Organic Matter Decomposition

1. **Decomposition of Plant Residues**:

- o **Breakdown of Organic Materials**: Mycorrhizal fungi, especially those in arbuscular mycorrhizal (AMF) and ectomycorrhizal (ECMF) groups, contribute to the decomposition of plant residues such as leaf litter, roots, and other organic materials. The fungal hyphae colonize decaying plant tissues and secrete enzymes that break down complex organic compounds into simpler forms. This process releases nutrients locked in organic matter, making them available for plant uptake.
- o **Interaction with Soil Microorganisms**: Mycorrhizal fungi interact with other soil microorganisms, including bacteria and actinomycetes, that also participate in decomposition. These interactions can enhance the efficiency of organic matter breakdown by synergistically working to decompose organic materials and mineralize nutrients.

2. **Formation of Humus**:

- o **Humification Process**: As organic matter decomposes, it is converted into humus, a stable form of organic matter that improves soil structure and fertility. Mycorrhizal fungi contribute to the humification process by breaking down organic matter into humic substances, which help to bind soil particles together and create stable soil aggregates.
- o **Soil Organic Matter (SOM)**: The humus formed through fungal decomposition becomes part of the soil organic matter pool. SOM is crucial for soil health as it improves soil water-holding capacity, enhances nutrient retention, and supports a diverse microbial community.

Benefits of Organic Matter Decomposition

1. **Nutrient Cycling**:

Release of Essential Nutrients: The decomposition of organic matter by mycorrhizal fungi releases essential nutrients such as nitrogen, phosphorus, potassium, and micronutrients into the soil. These nutrients are then available for plant uptake, supporting healthy plant growth and development.

Improved Fertility: The continuous decomposition of organic matter ensures a steady supply of nutrients to plants, improving soil fertility and reducing the need for synthetic fertilizers. This contributes to more sustainable agricultural practices and enhances soil productivity.

2. **Soil Structure and Health**:

Soil Aggregation: The formation of humus and the addition of organic matter to the soil contribute to the development of stable soil aggregates. These aggregates improve soil structure, enhance porosity, and increase water infiltration, reducing surface runoff and erosion.

Soil Moisture Retention: Organic matter helps to retain moisture in the soil by increasing its water-holding capacity. This is particularly important for maintaining soil moisture during dry periods and supporting plant growth.

Ecological and Agricultural Implications

1. **Ecosystem Health**:

   Support for Soil Biodiversity: The decomposition of organic matter by mycorrhizal fungi supports a diverse range of soil organisms, including bacteria, earthworms, and other fungi. This biodiversity is essential for maintaining healthy soil ecosystems and promoting ecological balance.

   Nutrient Availability: By facilitating nutrient cycling and improving soil fertility, mycorrhizal fungi contribute to the health and productivity of natural ecosystems. This supports plant communities and enhances ecosystem resilience.

2. **Sustainable Farming**:

   Reduced Soil Erosion: Well-aggregated soils with high organic matter content are less prone to erosion. Mycorrhizal fungi contribute to soil aggregation and stability, reducing the risk of erosion and maintaining soil health.

   Organic Farming Practices: Mycorrhizal inoculants are used in organic farming to enhance nutrient availability and support soil health. By improving the decomposition of organic materials and nutrient cycling, mycorrhizal fungi contribute to the success of organic farming systems.

Factors Influencing Decomposition

1. **Environmental Conditions**:

   Temperature and Moisture: Decomposition rates are influenced by environmental factors such as temperature and moisture. Optimal conditions for fungal activity include moderate temperatures and adequate soil moisture. Extreme conditions can slow down the decomposition process and affect nutrient availability.

   Soil pH: Soil pH can also impact the efficiency of decomposition. Most mycorrhizal fungi prefer slightly acidic to neutral pH levels, and extreme pH values can inhibit fungal activity and decomposition.

2. **Organic Matter Quality**:

Type of Organic Matter: The quality and composition of organic matter affect the rate of decomposition. Materials with high lignin or cellulose content, such as woody residues, decompose more slowly than those with higher nitrogen content. Mycorrhizal fungi can adapt to different types of organic matter and contribute to their breakdown.

Mycorrhizal fungi play a crucial role in organic matter decomposition, which is essential for maintaining soil health and fertility. Their activities contribute to nutrient cycling, soil structure, and overall ecosystem functioning. By enhancing the decomposition process, mycorrhizal fungi support sustainable agricultural practices and promote the health of natural soil ecosystems.

Enhancing Plant Health and Resilience

1. Disease and Pest Resistance: Mycorrhizal fungi can enhance plant resistance to soil-borne pathogens and pests. By colonizing the root zone, they form a physical barrier that reduces pathogen invasion. Additionally, they produce antimicrobial compounds that help protect plants from diseases. This is particularly important in reducing the need for chemical pesticides. Mycorrhizal fungi play a crucial role in enhancing plant resistance to diseases and pests, which is a significant factor in sustainable agriculture and ecosystem health. Their interactions with plant roots and the soil environment provide multiple mechanisms through which they bolster plant defenses and improve resilience against biotic stressors.

Mechanisms of Disease and Pest Resistance

1. **Enhanced Plant Immune Responses**:

   Induction of Systemic Resistance: Mycorrhizal fungi can trigger systemic resistance in plants, a defense mechanism that enhances the plant's ability to resist a broad spectrum of pathogens. When plants are colonized by mycorrhizal fungi, they often exhibit an increase in the production of defense-related compounds, such as phytoalexins, pathogenesis-related proteins, and antioxidant enzymes. These compounds help to strengthen the plant's immune system and improve its ability to fend off infections.

   Improved Physical Barriers: Mycorrhizal fungi contribute to the development of physical barriers in plant tissues, such as thicker cell walls and increased production of lignin and suberin. These structural defenses make it more difficult for pathogens to penetrate and cause disease.

2. **Competition and Antagonism**:

   Competitive Exclusion: Mycorrhizal fungi can outcompete pathogenic microorganisms for space and resources in the rhizosphere. By colonizing the root zone and forming a dense network of hyphae, they reduce the availability of niches and nutrients for pathogens, thereby limiting their ability to establish and proliferate.

Production of Antimicrobial Compounds: Some mycorrhizal fungi produce antimicrobial compounds that inhibit the growth of pathogens. These compounds can directly suppress the activity of harmful microorganisms or prevent their colonization by altering the microbial community in the soil.

Enhanced Plant Growth and Stress Tolerance

1. **Improved Nutrient and Water Uptake**:

   Nutrient Acquisition: Mycorrhizal fungi enhance the uptake of essential nutrients, such as phosphorus, nitrogen, and other minerals. Improved nutrient status helps plants to maintain robust growth and better withstand stress from pests and diseases. For example, plants with adequate phosphorus levels are less susceptible to certain fungal diseases.

   Water Stress Resistance: Mycorrhizal fungi improve soil water retention and increase plant drought tolerance. Well-watered and healthy plants are less prone to stress-induced diseases and are better able to resist pest attacks.

2. **Reduced Plant Stress**:

   Stress Resilience: Mycorrhizal fungi contribute to plant resilience by mitigating the effects of abiotic stresses, such as drought, salinity, and extreme temperatures. Plants that are less stressed are less likely to be vulnerable to opportunistic pathogens and pests.

Examples of Mycorrhizal Influence on Disease and Pest Resistance

1. **Arbuscular Mycorrhizal Fungi (AMF)**:

   Disease Suppression: AMF have been shown to provide resistance to various soil-borne pathogens, such as root rot fungi and nematodes. For instance, AMF colonization has been associated with reduced incidence of root rot caused by pathogens like *Fusarium* and *Phytophthora* species.

   Pest Resistance: AMF can also contribute to resistance against certain insect pests by improving plant health and growth. Plants colonized by AMF often exhibit increased resistance to pests such as aphids and root-feeding nematodes.

2. **Ectomycorrhizal Fungi (ECMF)**:

Enhanced Defense Mechanisms: ECMF are known to enhance the defense mechanisms of forest trees. For example, ECMF colonization in trees like oak and pine has been linked to increased production of defensive compounds, such as phenolics and terpenoids, which deter herbivores and pathogens.

Disease Tolerance: ECMF can also improve tolerance to foliar diseases by enhancing the overall health of the plant and strengthening its immune responses.

Practical Applications in Agriculture

1. **Integrated Pest Management (IPM)**:

   Mycorrhizal Inoculation: Incorporating mycorrhizal inoculants into agricultural practices can be part of an Integrated Pest Management (IPM) strategy. By enhancing plant health and resistance, mycorrhizal fungi can reduce the reliance on chemical pesticides and promote more sustainable farming practices.

   Crop Rotations and Soil Management: Practices that support the growth of mycorrhizal fungi, such as crop rotations and organic soil amendments, can improve soil health and reduce the incidence of pest and disease problems.

2. **Organic Farming**:

   Sustainable Practices: Mycorrhizal fungi are commonly used in organic farming systems to enhance plant health and resistance. Organic farmers often utilize mycorrhizal inoculants and maintain practices that support fungal growth, such as minimizing soil disturbance and using compost.

Future Research Directions

1. **Understanding Fungal Interactions**:

   Molecular Mechanisms: Further research is needed to understand the molecular mechanisms through which mycorrhizal fungi induce plant resistance. This includes identifying specific signaling pathways and defensive compounds involved in the mycorrhizal-plant interaction.

   Diverse Fungal Species: Investigating the effects of different mycorrhizal fungal species on various plant pathogens and pests will help to tailor mycorrhizal inoculation strategies for specific crops and environmental conditions.

2. **Field Studies and Applications**:

Real-World Applications: Large-scale field studies are required to evaluate the effectiveness of mycorrhizal inoculants in managing pests and diseases under diverse agricultural and ecological conditions. This will help to optimize the use of mycorrhizal fungi in practical applications.

Mycorrhizal fungi play a significant role in enhancing plant resistance to diseases and pests through improved immune responses, competition with pathogens, and support for plant growth. Their contributions to soil health and plant resilience make them valuable tools in sustainable agriculture and ecosystem management. Understanding and harnessing these benefits can lead to more effective and environmentally friendly strategies for managing plant health.

2. Stress Tolerance: Mycorrhizal fungi can improve plant tolerance to various environmental stresses, including drought, salinity, and heavy metal toxicity. They enhance the plant's ability to absorb water and nutrients, which helps plants cope with adverse conditions. For example, mycorrhizal associations can increase root hydration and reduce the impact of soil salinity on plant health. Mycorrhizal fungi are crucial for improving plant stress tolerance, particularly in the face of abiotic stresses such as drought, salinity, extreme temperatures, and soil acidity. Their interactions with plant roots enhance the plant's ability to withstand and recover from various stress conditions, leading to more resilient agricultural and natural ecosystems.

Mechanisms of Stress Tolerance Enhancement

1. **Improved Water and Nutrient Uptake**:

     Enhanced Water Absorption: Mycorrhizal fungi form extensive networks of hyphae that extend beyond the root zone, increasing the surface area for water absorption. This allows plants to access moisture from a larger soil volume, which is especially beneficial during drought conditions. By improving water uptake, mycorrhizal fungi help plants maintain hydration and reduce water stress.

     Nutrient Acquisition: Mycorrhizal fungi enhance the uptake of essential nutrients such as phosphorus, nitrogen, and potassium. Adequate nutrient supply helps plants maintain metabolic functions and cellular processes under stress conditions. For instance, phosphorus acquisition is particularly crucial for energy transfer and stress response mechanisms in plants.

2. **Reduction of Soil Stress Effects**:

     Soil Structure Improvement: Mycorrhizal fungi contribute to soil aggregation and improve soil structure by binding soil particles together. This results in better aeration and reduced soil compaction, which can alleviate root stress and enhance root growth. Improved soil structure also helps with water infiltration and retention, mitigating the effects of drought.

Soil pH and Toxicity Management: Mycorrhizal fungi can help manage soil pH and reduce the toxicity of certain soil elements. For example, they can assist in the uptake of toxic metals and help plants tolerate acidic or alkaline soils by altering the rhizosphere environment.

Specific Stress Tolerance Enhancements

1. **Drought Tolerance**:

   Water Use Efficiency: Mycorrhizal fungi enhance the plant's ability to utilize available water more efficiently. The fungal hyphae act as extensions of the plant root system, accessing deeper soil moisture and improving overall water uptake. This is especially beneficial during periods of low precipitation or extended dry spells.

   Osmotic Adjustment: Mycorrhizal fungi can also influence the plant's osmotic adjustment mechanisms by enhancing the synthesis of osmoprotectants, such as proline and glycine betaine, which help stabilize cellular structures and protect against dehydration.

2. **Salinity Tolerance**:

   Salt Exclusion and Accumulation: Mycorrhizal fungi can improve plant tolerance to salinity by influencing ion uptake and distribution. They help plants exclude or sequester excess salts in vacuoles, reducing the toxic effects of salt stress on cellular functions.

   Improved Root Function: By enhancing root growth and function, mycorrhizal fungi support better water and nutrient uptake, which can counteract some of the negative impacts of high salinity on plant health.

3. **Temperature Extremes**:

   Heat Stress: Mycorrhizal fungi can help plants cope with high temperatures by maintaining better water and nutrient status. They may also enhance the production of heat shock proteins that protect cellular structures from heat damage.

   Cold Stress: During cold conditions, mycorrhizal fungi support root health and nutrient uptake, helping plants to sustain metabolic processes and growth even in low temperatures.

Ecological and Agricultural Implications

1. **Ecosystem Resilience**:

   Natural Ecosystems: Mycorrhizal fungi contribute to the resilience of natural ecosystems by enhancing plant stress tolerance. In diverse environments, such as arid or temperate regions, their presence supports plant survival and ecosystem stability.

Biodiversity Conservation: By improving stress tolerance in plants, mycorrhizal fungi help maintain plant diversity and support the functioning of various ecosystems, which is critical for biodiversity conservation.

2. **Sustainable Agriculture**:

   Crop Management: Mycorrhizal inoculants are used in sustainable agriculture to enhance crop resilience to stress. This can reduce the need for synthetic inputs, such as fertilizers and pesticides, and support more environmentally friendly farming practices.

   Climate Change Adaptation: Mycorrhizal fungi can play a role in adapting agricultural systems to climate change by improving plant tolerance to extreme weather events and fluctuating environmental conditions.

Future Research Directions

1. **Mechanistic Understanding**:

   Molecular Pathways: Research is needed to elucidate the molecular pathways through which mycorrhizal fungi enhance stress tolerance. Understanding the specific genes and proteins involved in stress response can help in developing targeted strategies for crop improvement.

   Fungal Strain Variability: Investigating the variability among different mycorrhizal fungal strains in their ability to confer stress tolerance will help identify the most effective strains for specific stress conditions.

2. **Field Applications**:

   Large-Scale Trials: Field trials are necessary to assess the effectiveness of mycorrhizal inoculants under real-world conditions. These trials can provide insights into the practical benefits of mycorrhizal fungi for enhancing stress tolerance in different crops and environments.

Mycorrhizal fungi significantly enhance plant stress tolerance through improved water and nutrient uptake, better soil structure, and mechanisms that mitigate the effects of drought, salinity, and temperature extremes. Their role in supporting plant resilience is crucial for both natural ecosystems and agricultural systems, making them valuable partners in managing stress and ensuring sustainable plant health.

Supporting Ecosystem Functions

1. Biodiversity: Mycorrhizal fungi play a role in supporting plant biodiversity. By enhancing nutrient availability and plant health, they contribute to the establishment and growth of diverse plant species. This, in turn, supports a wide range of organisms within the ecosystem, including insects, fungi, and other microorganisms.

2. Carbon Sequestration: Mycorrhizal fungi contribute to soil carbon sequestration. The decomposition of organic matter by mycorrhizal fungi and the formation of stable soil organic matter (humus) result in the storage of carbon in the soil. This process helps mitigate climate change by reducing atmospheric carbon dioxide levels.

Agricultural Applications

1. Sustainable Farming: In agriculture, mycorrhizal inoculants are used to improve crop yields and reduce the dependence on synthetic fertilizers. By enhancing nutrient uptake and soil health, mycorrhizal fungi contribute to more sustainable and environmentally friendly farming practices.

2. Restoration and Rehabilitation: Mycorrhizal fungi are used in ecological restoration and land rehabilitation projects. They help establish vegetation in disturbed or degraded soils, improve soil fertility, and enhance plant survival in challenging conditions.

Mycorrhizae are integral to nutrient cycling, soil health, plant resilience, and ecosystem functioning. Their ability to enhance nutrient uptake, improve soil structure, and support plant health underscores their importance in maintaining and restoring ecological balance and productivity.

3.3.Fungal contribution to soil structure and stability.

Fungi play an essential role in soil structure and stability, significantly influencing the physical properties of soils through their interactions with organic matter, minerals, and plant roots. By forming symbiotic relationships with plants and decomposing organic material, fungi contribute to the development and maintenance of soil aggregates, which are critical for healthy soil functioning. The structural integrity of soil is a fundamental factor in water infiltration, nutrient retention, and overall soil fertility, making fungi key organisms in both natural ecosystems and agricultural environments.

Fungal Hyphae and Soil Aggregation

1. **Hyphal Networks**:

    Fungal Hyphae: The filamentous structures of fungi, known as hyphae, extend throughout the soil, forming an interconnected web that physically binds soil particles together. These hyphal networks are particularly effective at stabilizing soil aggregates, which are clusters of soil particles bound together by organic and inorganic substances.

Aggregate Formation: Soil aggregation is a process where smaller soil particles, such as clay, silt, and sand, are bound into larger, more stable clumps or aggregates. Fungal hyphae play a crucial role in this by acting as biological "threads" that weave through the soil, pulling together particles and holding them in place. This not only protects the soil from erosion but also enhances its structural integrity, allowing for better aeration and root penetration.

2. **Excretion of Polysaccharides**:

Glomalin Production: Mycorrhizal fungi, particularly arbuscular mycorrhizal fungi (AMF), secrete a protein called glomalin, which is a key factor in soil aggregation. Glomalin acts as a glue-like substance that binds soil particles and organic matter into stable aggregates. This contributes to the long-term stability of soil structure by creating aggregates that are resistant to physical breakdown, even during heavy rainfall or other forms of soil disturbance.

Polysaccharide Binding: In addition to glomalin, fungi produce extracellular polysaccharides and other organic compounds that help to cement soil particles together. These compounds form a sticky matrix that traps fine soil particles and organic matter, reinforcing the soil's physical structure.

Fungi and Soil Stability

1. **Resistance to Erosion**:

Soil Protection from Erosion: One of the primary benefits of fungi in soil is their role in protecting the soil from erosion. The hyphal networks created by fungi help anchor soil particles, reducing the risk of erosion caused by wind, water, or agricultural activities. By stabilizing soil aggregates, fungi prevent the detachment and transport of soil particles, which can lead to the degradation of topsoil, the most fertile layer of the soil.

Improved Water Infiltration: The presence of well-formed soil aggregates also improves the soil's ability to absorb and retain water. Fungi help create pore spaces within the soil structure, which enhances water infiltration and reduces surface runoff. This is particularly important in preventing soil erosion during heavy rainfall, as aggregated soils are less likely to become compacted or washed away.

2. **Soil Porosity and Root Penetration**:

Enhancing Soil Porosity: Fungal activity in the soil increases porosity, which refers to the spaces between soil particles. These spaces are critical for air and water movement through the soil profile. By maintaining an open, porous soil structure, fungi ensure that roots can access oxygen and nutrients more easily, promoting healthy plant growth. Improved soil porosity also helps mitigate problems such as waterlogging or poor drainage, which can negatively impact soil health.

Facilitating Root Growth: The hyphal networks of fungi can penetrate even compact or dense soils, creating channels that facilitate the movement of plant roots. This is particularly beneficial in subsoils where compaction can limit root penetration. By loosening the soil and creating micro-pores, fungi enhance root development, allowing plants to access deeper soil layers and increasing their ability to extract water and nutrients.

Fungi and Soil Organic Matter

1. **Decomposition of Organic Matter**:

   Fungi as Decomposers: Saprotrophic fungi are key decomposers in the soil ecosystem, breaking down complex organic materials such as plant litter, dead organisms, and animal waste into simpler compounds. This decomposition process contributes to the formation of humus, a stable form of organic matter that plays a vital role in soil structure and fertility.

   Contribution to Soil Organic Carbon: Fungal decomposition contributes to the accumulation of soil organic carbon, which is a crucial component of soil health. Organic carbon helps bind soil particles together, enhancing soil aggregate stability. It also improves the soil's ability to retain moisture and nutrients, creating a more favorable environment for plant growth and microbial activity.

2. **Humus Formation and Stability**:

   Humus and Soil Aggregation: The humus formed through fungal decomposition is highly stable and plays a key role in maintaining soil structure. It binds to mineral particles and organic residues, forming larger aggregates that enhance soil stability and fertility. Humus also improves the soil's water-holding capacity, helping to buffer the soil against drought and nutrient loss.

   Long-Term Carbon Sequestration: Fungal-mediated decomposition also contributes to long-term carbon storage in the soil. By incorporating carbon into stable humus, fungi help sequester atmospheric carbon dioxide, which is critical for mitigating climate change. This stored carbon is less likely to be rapidly decomposed, allowing it to remain in the soil for extended periods, further stabilizing the soil structure.

Mycorrhizal Fungi and Soil Health

1. **Arbuscular Mycorrhizal Fungi (AMF)**:

Root-Soil Interface: AMF form symbiotic relationships with plant roots, improving nutrient uptake and contributing to soil structure. The extensive hyphal networks of AMF connect plant roots to a larger soil volume, enhancing nutrient and water absorption. In addition, AMF hyphae bind soil particles together, promoting soil aggregation and reducing the risk of erosion.

Glomalin and Soil Aggregation: The production of glomalin by AMF is a critical factor in maintaining stable soil aggregates. Glomalin improves the physical structure of soil, making it more resistant to erosion, compaction, and nutrient leaching. This, in turn, creates a healthier soil environment for both plants and microorganisms.

2. **Ectomycorrhizal Fungi (ECM)**:

Forest Soil Stability: In forest ecosystems, ECM play a similar role in stabilizing soil structure by forming symbiotic relationships with tree roots. ECM help trees access nutrients and water while also contributing to soil aggregation. Their hyphal networks create channels in the soil that facilitate root growth and enhance soil porosity, improving the overall health and stability of forest soils.

Organic Matter Decomposition: ECM fungi also contribute to the decomposition of organic matter, particularly in forest soils where they help break down leaf litter and other plant residues. This process not only recycles nutrients but also contributes to the formation of stable soil aggregates that protect the soil from erosion and degradation.

Fungi are integral to soil structure and stability through their contributions to soil aggregation, organic matter decomposition, and nutrient cycling. By binding soil particles together, improving soil porosity, and protecting against erosion, fungi ensure the long-term health and productivity of soils. Their roles in both natural ecosystems and agricultural environments make them essential for sustainable land management practices and ecological resilience.

3.4. Conclusion

Soil fungi are vital components of ecosystems, serving as fundamental drivers of key processes that sustain life on Earth. Their contributions to nutrient cycling, organic matter decomposition, soil structure formation, and symbiotic relationships with plants underscore their essential role in maintaining ecosystem health and productivity. As decomposers, fungi break down complex organic matter, releasing nutrients like carbon, nitrogen, and phosphorus back into the soil, which plants and other organisms rely on for growth. In symbiosis with plants, mycorrhizal fungi enhance nutrient uptake, particularly phosphorus, while also improving plant resilience to stress. These interactions contribute to the biodiversity and functionality of ecosystems, ensuring that nutrients are efficiently cycled and that plants thrive even under challenging environmental conditions. Fungal activity also promotes soil health by enhancing soil structure, increasing porosity, and improving water retention, which collectively result in more stable and fertile soils.

In addition to their ecological roles, soil fungi are becoming increasingly recognized for their potential in sustainable agriculture and environmental conservation. The understanding of fungal contributions to soil stability and ecosystem resilience can inform practices such as mycorrhizal inoculation in agriculture to improve crop yields, reduce dependency on chemical fertilizers, and enhance soil sustainability. By recognizing fungi's multifaceted roles—from nutrient cycling to plant symbiosis and soil health enhancement—scientists, ecologists, and farmers can harness these organisms to promote healthier ecosystems and more sustainable food systems. As ecosystems face increasing pressures from climate change, habitat destruction, and soil degradation, soil fungi offer a promising natural solution to enhance ecosystem functions and support long-term environmental resilience

Chapter 4

Fungal Interactions with Plants

Fungi form a diverse range of interactions with plants, which can range from mutually beneficial symbioses to parasitic relationships. One of the most significant and ecologically important types of fungal-plant interaction is the symbiotic relationship known as mycorrhizae. In this partnership, fungi colonize plant roots, providing plants with enhanced access to nutrients, particularly phosphorus, in exchange for carbohydrates that plants produce via photosynthesis. Mycorrhizal fungi extend the root system of the plant by forming vast networks of hyphae in the soil, which increases the plant's ability to absorb water and essential nutrients, such as nitrogen and phosphorus. This symbiosis improves plant growth and helps plants survive in nutrient-poor environments. In return, the fungi receive sugars and other organic compounds that the plant synthesizes, establishing a mutual exchange that supports both partners.

Aside from nutrient acquisition, fungal interactions with plants also help in stress mitigation. Mycorrhizal fungi can enhance a plant's tolerance to environmental stresses such as drought, salinity, and heavy metal toxicity. Additionally, mycorrhizal associations can protect plants from root pathogens by forming a physical barrier and by stimulating the plant's immune response. Endophytic fungi also inhabit plant tissues without causing harm and may even confer benefits such as increased resistance to diseases and herbivores. These endophytes can produce bioactive compounds that deter pathogens and pests, contributing to the overall health and longevity of the plant. Parasitic fungi, on the other hand, can harm plants by causing diseases, such as rusts, smuts, and blights. These interactions illustrate the complexity and importance of fungal-plant relationships, as they play key roles in ecosystem stability, plant health, and agricultural productivity.

4.1. How soil fungi support plant growth and health.

Soil fungi are crucial allies in supporting plant growth and health, largely through their role in nutrient acquisition, particularly in nutrient-poor environments. One of the most significant ways fungi support plants is via mycorrhizal associations. Mycorrhizal fungi colonize plant roots, extending the root system by forming an extensive network of hyphae that increases the surface area for nutrient absorption. This fungal network allows plants to access nutrients like phosphorus, nitrogen, and other trace elements that would otherwise be unavailable due to their low mobility in soil. By facilitating better nutrient uptake, these fungi improve plant growth, yield, and overall vitality. In exchange, the plant provides the fungi with carbohydrates derived from photosynthesis, creating a symbiotic relationship that benefits both organisms. In poor or degraded soils, these relationships become even more critical, enabling plants to thrive in otherwise challenging conditions. Beyond nutrient acquisition, soil fungi also play a vital role in enhancing plant health by boosting resilience against environmental stressors and pathogens. For instance, mycorrhizal fungi improve a plant's ability to withstand drought by enhancing water uptake and improving the soil's water retention capacity. These fungi also help plants cope with salinity, heavy metal contamination, and temperature extremes by altering root function and improving stress tolerance. Additionally, fungi can act as biological control agents, protecting plants from root pathogens and harmful microorganisms. Some fungi form a protective barrier around plant roots or stimulate the plant's immune system, reducing the likelihood of infection by parasitic fungi and other pathogens. This combination of improved nutrient access, stress resilience, and disease protection underscores the pivotal role soil fungi play in promoting healthy, productive plant ecosystems.

Soil fungi play a critical role in supporting plant growth and health through their unique abilities to facilitate nutrient acquisition, enhance water uptake, and protect plants from both abiotic and biotic stressors. One of the most significant fungal roles is through **mycorrhizal associations**, in which fungi colonize plant roots and extend their hyphal networks deep into the soil. These networks act as an extension of the root system, significantly increasing the surface area available for nutrient absorption. Mycorrhizal fungi are especially effective at accessing immobile nutrients like phosphorus, nitrogen, and micronutrients such as zinc and copper. These nutrients are often bound in forms that are difficult for plants to access on their own, but the fungi can break them down and transport them to the plant. This symbiotic relationship is critical in poor soils, where nutrient availability is low. In return, plants provide the fungi with carbon-rich compounds generated during photosynthesis, creating a mutually beneficial exchange that enhances plant growth, development, and yield.

In addition to nutrient acquisition, soil fungi significantly contribute to improving plant resilience against environmental stresses. Mycorrhizal fungi improve the plant's ability to uptake water, especially during drought conditions, by accessing water reserves that lie beyond the reach of plant roots. This enhanced water absorption capability helps plants survive in dry or arid conditions and improves their overall drought tolerance. Furthermore, fungi can aid in mitigating the harmful effects of soil salinity and heavy metal toxicity, as their hyphal networks buffer the plant roots from absorbing excessive amounts of toxic substances. On the biological front, fungi help defend plants against pathogens and diseases. Some fungi form a physical barrier around plant roots, shielding them from harmful microbes and soil-borne diseases. Additionally, mycorrhizal fungi can stimulate the plant's immune system, enhancing its defense responses against attacks by parasitic organisms. This dual role of nutrient provision and pathogen protection makes soil fungi indispensable for plant health, productivity, and ecological resilience. Soil fungi play a pivotal role in supporting plant growth and health through several ecological functions, including nutrient acquisition, improving soil structure, promoting plant resilience to environmental stressors, and enhancing defense mechanisms against pathogens. The symbiotic relationships between fungi and plants, particularly mycorrhizal associations, are among the most important mechanisms through which fungi contribute to plant health.

Nutrient Acquisition:

One of the most critical ways fungi support plant growths is by enhancing the uptake of essential nutrients, particularly phosphorus, nitrogen, and trace elements like zinc and copper. This is primarily facilitated by mycorrhizal fungi, which form mutualistic relationships with plant roots. The two main types of mycorrhizal fungi are arbuscular mycorrhizal fungi (AMF) and ectomycorrhizal fungi, both of which extend their hyphal networks far beyond the root zone, exploring the soil for nutrients that plants cannot access on their own.

Arbuscular mycorrhizal fungi (AMF): AMF penetrate the root cells of most land plants and form specialized structures known as arbuscules, which facilitate the exchange of nutrients. AMF absorb inorganic phosphorus from the soil and deliver it to the plant in exchange for carbohydrates (sugars) that the plant produces through photosynthesis. In phosphorus-deficient soils, this symbiotic relationship is crucial for plant growth.

Ectomycorrhizal fungi: Ectomycorrhizal fungi primarily associate with woody plants, including trees like pine, oak, and birch. These fungi form a sheath around the plant's root tips and extend their hyphae into the surrounding soil. Like AMF, they enhance phosphorus and nitrogen uptake but are particularly effective in acquiring nitrogen in organic forms, as they can break down organic material.

These fungi also assist in the uptake of less mobile nutrients, such as zinc and copper, which are vital for various plant physiological processes, including enzyme function and photosynthesis.

Improving Soil Structure:

Fungi, particularly those that form extensive hyphal networks in the soil, contribute significantly to improving soil structure. Their hyphae act as physical binders, aggregating soil particles and organic matter into stable structures known as soil aggregates. These aggregates enhance the soil's porosity, allowing for better aeration, water infiltration, and root penetration. A well-aggregated soil holds more water and nutrients, which are readily available for plants, promoting their growth.

Moreover, fungi produce organic compounds like glomalin, a sticky glycoprotein secreted by AMF, which acts as a glue to bind soil particles together. Glomalin contributes to the formation of stable soil aggregates, further improving soil structure, water retention, and resistance to erosion. This improved soil structure also enhances root growth and nutrient absorption, providing a better environment for plants to thrive.

Increased Resistance to Environmental Stressors:

Soil fungi also enhance plant resilience to various environmental stressors, such as drought, salinity, and heavy metal toxicity. Mycorrhizal fungi, in particular, can help plants withstand these harsh conditions by improving water and nutrient uptake and modifying plant root architecture to be more efficient in resource acquisition.

Drought tolerance: Fungal hyphae can reach deep into the soil, accessing water that plant roots alone may not be able to reach. In drought-prone environments, plants with mycorrhizal associations have been shown to have higher water-use efficiency. The extensive fungal network helps the plant maintain hydration and continue metabolic processes even when water availability is low.

Salt stress: In saline soils, mycorrhizal fungi reduce the accumulation of toxic ions, such as sodium, in plant tissues. They help in maintaining a balance of essential nutrients like potassium and calcium, which are critical for maintaining cellular functions in salt-stressed plants.

Heavy metal detoxification: Some fungi can also detoxify soils contaminated with heavy metals, such as cadmium, lead, or arsenic, by binding these metals to their hyphal walls or converting them into less toxic forms. This reduces the uptake of harmful metals by plants, protecting them from toxic effects.

Enhanced Plant Defense Mechanisms:

Soil fungi also play a role in enhancing plant defense mechanisms against pathogens. Mycorrhizal fungi, through their physical presence and biochemical interactions with the host plant, can trigger systemic acquired resistance (SAR), a plant-wide immune response that makes the plant more resistant to pathogens and pests.

Biological control: Some soil fungi, such as Trichoderma species, are known for their antagonistic effects on plant pathogens. They can outcompete harmful microorganisms for space and nutrients or directly attack them by producing enzymes that degrade pathogen cell walls. For instance, Trichoderma fungi produce chitinase and glucanase enzymes, which break down the cell walls of fungal pathogens, protecting the host plant.

Induced resistance: Mycorrhizal fungi can induce physiological changes in the plant, including the production of defensive compounds like phenolics and phytoalexins. These compounds make the plant more resistant to microbial attacks by strengthening cell walls and inhibiting the growth of pathogens. This form of mycorrhiza-induced resistance is important in reducing the severity of diseases caused by soil-borne pathogens, such as root rot and wilt.

Promotion of Plant Growth Hormones:

Fungi can produce and influence the production of plant growth hormones, such as auxins, gibberellins, and cytokinins, which regulate various aspects of plant growth, including root development, shoot elongation, and flowering. For instance, AMF have been shown to stimulate root growth by enhancing the production of auxins, which increases the surface area of the roots for better nutrient and water absorption. This hormonal interaction promotes overall plant health, allowing for more robust growth even in nutrient-poor soils.

Supporting Plant Biodiversity:

Fungal diversity in soil fosters plant biodiversity by reducing competition for resources and enhancing nutrient availability in ecosystems. Mycorrhizal networks, in particular, create interconnected systems between plants, facilitating nutrient transfer and communication between them. This can allow less competitive plants to thrive in nutrient-limited environments, promoting greater diversity in plant communities.

In summary, soil fungi support plant growth and health by enhancing nutrient and water uptake, improving soil structure, increasing resilience to environmental stressors, boosting defense mechanisms, and promoting plant growth hormones. These diverse functions highlight fungi's critical role in maintaining healthy, productive ecosystems, and sustaining plant life across various environments.

4.2. The mycorrhizal network: How fungi connect plants through underground networks.

The mycorrhizal network refers to the intricate underground symbiotic system formed between plant roots and certain types of fungi. This network is one of nature's most efficient and widespread cooperative systems, benefiting both plants and fungi. The word "mycorrhiza" itself comes from Greek, meaning "fungus" (mykes) and "root" (rhiza), symbolizing this close partnership. In a mycorrhizal network, fungi colonize plant roots and form extensive hyphal networks in the soil, which extend far beyond the reach of the plant's natural root system. These fungal hyphae, which are fine, thread-like structures, explore a much greater volume of soil than roots can alone, allowing for increased absorption of water and nutrients, particularly phosphorus, nitrogen, and trace minerals that are often scarce or immobile in the soil. In return, the plants provide the fungi with carbohydrates and sugars, produced through photosynthesis, which the fungi cannot generate on their own. This mutually beneficial exchange strengthens the growth and survival of both partners, forming the basis for widespread ecological cooperation in terrestrial ecosystems.

There are two main types of mycorrhizal networks: arbuscular mycorrhizal fungi (AMF) and ectomycorrhizal fungi (EMF). Arbuscular mycorrhizal fungi penetrate the root cells of plants, forming arbuscules, which are specialized structures that facilitate the exchange of nutrients between the plant and fungi. AMF are incredibly common and are found in association with about 80% of plant species, particularly in herbaceous plants, grasses, and crops. Ectomycorrhizal fungi, on the other hand, surround the root tips with a sheath of hyphae but do not penetrate root cells. They form a network known as the Hartig net, which allows the exchange of nutrients outside the root cells. Ectomycorrhizal fungi are primarily associated with woody plants, such as trees in boreal and temperate forests, and they play an essential role in forest ecosystems. The mycorrhizal network not only enhances nutrient and water uptake but also connects different plants through a shared fungal network, sometimes referred to as the **"Wood Wide Web."** Through this network, resources can be transferred between plants, enabling stronger plants to share nutrients with weaker or younger plants. This communal exchange enhances the resilience and health of entire ecosystems, highlighting the crucial role of mycorrhizal networks in maintaining biodiversity, promoting plant health, and improving soil structure.

Fungi connect plants through an intricate and widespread underground system known as the mycorrhizal network, sometimes called the "Wood Wide Web." This network is formed by fungal hyphae—long, thread-like structures that extend into the soil from the roots of plants. When fungi colonize plant roots, particularly through mycorrhizal associations, they create connections between individual plants, allowing them to interact and share resources through this subterranean system. These networks can link multiple plants across vast areas, forming a communal network that enhances nutrient and water uptake while facilitating interplant communication and resource sharing.

The key to this underground connection is the symbiotic relationship between mycorrhizal fungi and plant roots. In arbuscular mycorrhizal fungi (AMF), the fungal hyphae penetrate the plant root cells and create specialized structures called arbuscules, where nutrient exchange occurs. Meanwhile, ectomycorrhizal fungi (EMF), which are common in forests, form a sheath around the root tips, and their hyphae grow outward into the soil, interconnecting with other plants. Through this network, fungi can facilitate the transfer of vital nutrients such as phosphorus, nitrogen, and carbon. The mycorrhizal network acts as a vast pipeline, redistributing resources, sometimes from larger, more established plants to smaller or younger ones. This sharing of nutrients enhances the growth and survival chances of weaker plants in the network, particularly under stressful conditions like nutrient-poor soils or drought.

This fungal network also plays an essential role in interplant communication. When a plant faces an attack, such as from pathogens or herbivores, it can send signals through the mycorrhizal network to neighboring plants. These signals, often in the form of chemical or hormonal cues, can activate defense mechanisms in the surrounding plants, preparing them for similar threats. Research has shown that plants connected through fungal networks exhibit heightened resistance to pathogens or herbivores compared to isolated plants. Additionally, the mycorrhizal network contributes to ecosystem stability by promoting plant diversity and

73

facilitating cooperation between species. Through this underground system, not only are individual plants able to grow more efficiently, but entire plant communities become more resilient, interconnected, and better equipped to withstand environmental stresses. The mycorrhizal network, often referred to as the "Wood Wide Web," is an intricate underground system formed by mycorrhizal fungi that connects plant roots across vast distances in ecosystems. This network serves as a conduit for the exchange of nutrients, water, and signaling compounds between plants, playing a crucial role in plant health, growth, and communication within plant communities. The mycorrhizal network is fundamental to plant survival in many ecosystems, especially forests, grasslands, and agricultural landscapes.

The Formation of Mycorrhizal Networks:

Mycorrhizal fungi form symbiotic relationships with plant roots, wherein the fungi colonize plant root tissues and extend their hyphae—thread-like structures—into the surrounding soil. There are two main types of mycorrhizal fungi involved in network formation: arbuscular mycorrhizal fungi (AMF) and ectomycorrhizal fungi (EMF).

> **Arbuscular mycorrhizal fungi (AMF):** AMF are the most widespread type, associating with around 80% of all plant species, including most grasses, herbs, and crop plants. AMF hyphae penetrate plant root cells and form specialized structures called arbuscules, where nutrient exchange between the plant and fungus occurs. These hyphae also extend out into the soil, connecting with other plant roots, creating a shared network.

> **Ectomycorrhizal fungi (EMF):** EMF primarily associate with trees and shrubs in temperate and boreal forests. Unlike AMF, EMF do not penetrate plant root cells but instead form a sheath around the root tips and grow into the intercellular spaces between root cells. Their hyphae extend into the soil and connect multiple trees and plants within the forest, often forming vast networks.

Once these fungi colonize plant roots, their hyphal networks link with other plants, connecting them together. This allows plants to share resources like water, nutrients, and signaling molecules through these underground fungal highways.

Nutrient Exchange through the Network:

One of the primary functions of the mycorrhizal network is the transfer of essential nutrients between plants. Mycorrhizal fungi have a remarkable ability to access soil nutrients that are otherwise unavailable to plants, such as phosphorus, nitrogen, potassium, and trace elements like zinc and copper. The fungi absorb these nutrients from the soil and transport them through their hyphal networks to connected plants.

Phosphorus and nitrogen: Phosphorus, a crucial nutrient for plant energy transfer and root development, is often present in insoluble forms in the soil, making it difficult for plants to absorb. Mycorrhizal fungi, particularly AMF, can solubilize and transport phosphorus directly to plant roots. Similarly, ectomycorrhizal fungi are adept at accessing organic nitrogen from the soil, breaking down organic matter, and supplying plants with this vital nutrient. Through the mycorrhizal network, plants can share excess nutrients with neighboring plants that might be nutrient-deficient.

Nutrient sharing among plants: In forests and grasslands, plants connected by the mycorrhizal network can "trade" nutrients. For example, older trees or plants with access to deep soil nutrients may supply younger or shaded plants with phosphorus or nitrogen in exchange for carbon compounds. This nutrient exchange helps maintain plant community health and stability, especially in ecosystems where nutrient availability is patchy or limited.

Water Transfer and Drought Tolerance:

The mycorrhizal network also facilitates the sharing of water between plants. In times of drought, water can be transferred from plants with greater access to water resources, such as deep-rooted trees, to plants experiencing water stress. This process helps improve drought tolerance for the entire plant community by redistributing water through the fungal network.

The extensive reach of mycorrhizal hyphae into the soil allows the network to access water from deeper or less accessible parts of the soil profile. This water is then made available to connected plants, enhancing their ability to survive and grow in arid conditions. The redistribution of water through the mycorrhizal network provides a buffer against drought and environmental stress, improving ecosystem resilience.

Communication and Defense Signaling:

The mycorrhizal network is not just a pathway for nutrients and water; it also serves as a communication system for plants. When plants are attacked by pathogens, herbivores, or experience other forms of stress, they release signaling compounds, such as defense hormones (e.g., jasmonic acid, salicylic acid) and secondary metabolites, into the mycorrhizal network. These signals are transmitted to neighboring plants through the fungal hyphae, allowing connected plants to "prepare" for potential threats.

Defense priming: Upon receiving signals from the mycorrhizal network, neighboring plants can ramp up their defenses by producing compounds like alkaloids, phenolics, or volatile organic compounds that deter herbivores or inhibit pathogen growth. This process is known as defense priming, where plants become more resistant to future attacks.

Communication beyond species boundaries: Interestingly, the mycorrhizal network can facilitate communication not only between individuals of the same species but also between different plant species. This cross-species communication can strengthen the entire plant community's resilience against pests and diseases.

Carbon Sharing and Resource Redistribution:

One of the most intriguing functions of the mycorrhizal network is the redistribution of carbon between plants. Plants produce carbohydrates through photosynthesis, some of which are shared with mycorrhizal fungi in exchange for nutrients. In ecosystems like forests, large trees—sometimes referred to as "mother trees"—can provide surplus carbon to smaller, shaded seedlings that are less capable of photosynthesis due to limited access to sunlight.

Mother trees and seedlings: Through the mycorrhizal network, carbon compounds move from older, well-established trees to young seedlings, helping them survive in low-light conditions under the forest canopy. This carbon sharing is particularly important in dense forests, where competition for light can be intense, and younger plants may struggle to survive without this support.

Resource redistribution: In addition to carbon, the mycorrhizal network helps redistribute other resources, such as nutrients and water, among plants in an ecosystem. This redistribution supports plant diversity by allowing less competitive species to thrive alongside dominant plants.

Influence on Plant Biodiversity and Ecosystem Stability:

The mycorrhizal network plays a critical role in shaping plant community dynamics and promoting biodiversity. By facilitating resource sharing and communication, the network reduces competition between plants and allows species with different resource requirements to coexist. This mutual support system helps maintain plant diversity in ecosystems like forests and grasslands.

Enhanced plant survival: The ability of plants to exchange resources through the mycorrhizal network enhances their chances of survival in challenging environments. This is especially important in nutrient-poor soils, where the network acts as a buffer against resource scarcity.

Ecosystem resilience: The mycorrhizal network enhances ecosystem resilience by promoting cooperation rather than competition among plants. In ecosystems prone to disturbances, such as fire, drought, or disease outbreaks, the network helps plants recover more quickly by providing access to shared resources, thereby stabilizing the ecosystem.

Disruption of Mycorrhizal Networks:

Despite their benefits, mycorrhizal networks are vulnerable to disruption, particularly due to human activities such as deforestation, agricultural practices, and the use of chemical fertilizers and pesticides.

These activities can reduce fungal diversity, degrade the network, and impair its ability to support plant communities. Intensive farming, for example, often involves tilling, which breaks up the fungal hyphae, and the application of high levels of phosphorus fertilizers, which can reduce the need for plants to engage in mycorrhizal associations.

Efforts to restore and conserve natural ecosystems increasingly recognize the importance of mycorrhizal networks in maintaining soil health and supporting plant biodiversity. Practices such as reforestation, agroforestry, and the use of organic farming techniques can help preserve these networks and the critical ecosystem services they provide.

The mycorrhizal network is a vital underground system that connects plants, allowing them to share nutrients, water, and signals in ways that enhance plant health, promote biodiversity, and improve ecosystem resilience. By facilitating cooperation among plants, these fungal networks support the survival and growth of entire plant communities, making them a cornerstone of ecological stability and functioning.

4.3.Fungi's role in plant immunity and resistance to pathogens.

Fungi play a crucial role in enhancing plant immunity and resistance to pathogens, primarily through symbiotic relationships like mycorrhizae and endophytic associations. Mycorrhizal fungi, particularly arbuscular mycorrhizal fungi (AMF) and ectomycorrhizal fungi (EMF), not only facilitate nutrient acquisition but also act as vital components of a plant's immune defense. These fungi form physical barriers around plant roots, preventing the direct invasion of soil-borne pathogens such as harmful fungi, bacteria, and nematodes. The mycorrhizal hyphae shield the plant's roots by creating a protective layer, limiting the ability of pathogens to colonize or penetrate plant tissues. In addition to this physical barrier, mycorrhizal fungi stimulate the plant's immune system through a process called systemic acquired resistance (SAR) and induced systemic resistance (ISR). When a plant is colonized by mycorrhizal fungi, the fungi trigger the production of signaling molecules, such as jasmonic acid and salicylic acid, that enhance the plant's innate immune response, priming it to better resist future pathogen attacks. This priming effect equips plants with stronger defense mechanisms, allowing them to respond more effectively to stress or infections.

Beyond the direct physical and chemical defenses, endophytic fungi—fungi that live inside plant tissues without causing harm—also contribute to plant immunity. These fungi often produce bioactive compounds, such as alkaloids, terpenoids, and phenolics, which are toxic or inhibitory to a variety of pathogens, including bacteria, fungi, and viruses. Some endophytes can also produce secondary metabolites that deter herbivores and reduce damage from pests, further enhancing the plant's ability to resist biotic stressors. Additionally, endophytic fungi can outcompete pathogenic organisms for space and resources within the plant tissues, limiting the spread and severity of infections. Fungi not only activate the plant's defense pathways but also help maintain a healthy microbial balance in the root and soil environment. By modulating the plant's immune response and directly producing antimicrobial compounds, fungi serve as a crucial line of defense, ensuring that plants remain resilient against pathogens and other environmental stresses. This symbiotic interaction between plants and fungi is fundamental to maintaining ecosystem health and agricultural productivity.

Fungi play a pivotal role in bolstering plant immunity and enhancing resistance to pathogens, primarily through symbiotic relationships such as mycorrhizae and endophytic associations. One of the most significant ways fungi contribute to plant defense is through the formation of mycorrhizal networks, which connect plant roots to extensive fungal hyphal systems. Mycorrhizal fungi, such as arbuscular mycorrhizal fungi (AMF) and ectomycorrhizal fungi (EMF), form protective layers around plant roots. This physical barrier acts as the first line of defense, preventing the penetration of harmful pathogens, such as parasitic fungi, bacteria, and nematodes. By limiting pathogen access to the plant's root system, these fungi reduce the likelihood of infection. Moreover, mycorrhizal fungi help suppress pathogens by competing with them for space and nutrients in the rhizosphere, thus creating a more hostile environment for potential invaders.

In addition to forming physical barriers, mycorrhizal fungi also enhance plant immune responses through biochemical signaling pathways. These fungi induce systemic changes in plants that activate their natural defense mechanisms. The presence of mycorrhizal fungi triggers systemic acquired resistance (SAR) and induced systemic resistance (ISR). These processes stimulate the production of defense-related compounds such as phytoalexins and pathogenesis-related proteins (PR proteins), which strengthen the plant's overall immunity. Mycorrhizal fungi also influence hormone pathways, particularly those involving jasmonic acid and salicylic acid, which are critical for regulating plant defense responses. By "priming" the plant's immune system, mycorrhizal fungi make the plant more prepared to respond rapidly and effectively to subsequent attacks from pathogens. This priming ensures that when a plant encounters a pathogen, it can mount a stronger and faster defense response, reducing the severity of infections.

Endophytic fungi, which reside inside plant tissues without causing harm, further contribute to plant defense. These fungi often produce secondary metabolites, such as alkaloids, terpenoids, and phenolic compounds, which are toxic to pathogens and herbivores. These antimicrobial compounds inhibit the growth and spread of invading pathogens, protecting plants from a wide range of diseases. Endophytic fungi also compete with pathogenic microorganisms within plant tissues, limiting their ability to colonize and cause damage. This biological competition, combined with the production of defensive chemicals, enhances the plant's overall resistance to both biotic and abiotic stresses. Additionally, some endophytic fungi can activate the plant's own production of defense compounds, further boosting its ability to withstand infections. Through these multifaceted mechanisms—physical protection, immune system activation, and biochemical defenses—fungi play a crucial role in helping plants resist pathogens and maintain health, thus contributing to the stability and productivity of ecosystems. Here's an in-depth look at how fungi contribute to plant immunity and pathogen resistance.

Mycorrhizal-Induced Resistance (MIR):

One of the primary ways fungi enhance plant immunity is through mycorrhizal-induced resistance (MIR), a form of systemic acquired resistance (SAR) that is triggered when plants form symbiotic relationships with mycorrhizal fungi. When plants engage with mycorrhizal fungi, the fungi colonize their roots, either penetrating the cells (in the case of arbuscular mycorrhizal fungi) or forming a sheath around the root (in the case of ectomycorrhizal fungi). This interaction initiates a cascade of immune responses that prepare the plant to defend itself against potential pathogen attacks.

Activation of plant defense pathways: The colonization of plant roots by mycorrhizal fungi leads to the activation of plant defense signaling pathways, particularly those involving salicylic acid (SA), jasmonic acid (JA), and ethylene (ET). These pathways are crucial for regulating the plant's immune responses to different types of pathogens. For example, salicylic acid is often involved in defending against biotrophic pathogens (which feed on living plant tissue), while jasmonic acid and ethylene are more active in responses against necrotrophic pathogens (which kill plant tissues and then feed on the dead matter).

Priming of plant defenses: Mycorrhizal fungi not only activate immune responses but also prime the plant's defense systems, a process called defense priming. Primed plants are able to respond more quickly and strongly when they encounter a pathogen. The plant doesn't continuously produce defensive compounds, which would be metabolically expensive, but instead remains in a heightened state of readiness. When a pathogen attacks, the primed plant rapidly ramps up its production of defensive chemicals, such as phenolics and phytoalexins, which inhibit the growth of pathogens.

Physical and Biochemical Changes in Plant Tissues:

Fungi can also induce physical and biochemical changes in plant tissues that strengthen their defenses against pathogens.

Thicker cell walls: Mycorrhizal fungi can stimulate plants to produce compounds like lignin and callose, which reinforce the plant's cell walls, making it more difficult for pathogens to penetrate the cells. These structural changes act as physical barriers to infection, particularly against pathogens that try to enter through wounds or natural openings in the plant.

Production of defensive proteins and enzymes: Plants associated with mycorrhizal fungi often increase the production of defense-related proteins, such as pathogenesis-related (PR) proteins, which have antimicrobial properties. For example, enzymes like chitinases and glucanases break down the cell walls of fungal pathogens, limiting their ability to invade plant tissues. Similarly, the production of reactive oxygen species (ROS), which can damage pathogen cells, is often upregulated in plants that are part of a mycorrhizal network.

Increased production of secondary metabolites: Mycorrhizal fungi can also enhance the production of secondary metabolites in plants, such as alkaloids, terpenoids, and phenolic compounds, which serve as chemical defenses against pathogens and herbivores. These compounds can have direct antimicrobial activity, inhibiting the growth of fungi, bacteria, and viruses that attack the plant.

Biocontrol of Pathogens through Antagonistic Fungi:

Certain soil fungi, such as *Trichoderma* and *Gliocladium* species, act as biocontrol agents by directly antagonizing plant pathogens. These fungi compete with pathogens for space and resources, produce antimicrobial compounds, and sometimes parasitize the pathogens themselves.

Competition for resources and space: Antagonistic fungi like Trichoderma grow rapidly in the rhizosphere (the area of soil surrounding plant roots), where they compete with pathogenic fungi for nutrients and space. By outcompeting harmful pathogens, these beneficial fungi reduce the chances of infection and disease in plants.

Production of antimicrobial compounds: Trichoderma and other biocontrol fungi secrete a variety of antimicrobial metabolites, including antibiotics, volatile organic compounds, and enzymes. These compounds can inhibit or kill pathogenic fungi, bacteria, and even nematodes in the soil. For example, Trichoderma species produce cellulase and chitinase, enzymes that degrade the cell walls of many fungal pathogens, such as *Rhizoctonia, Pythium,* and *Fusarium.*

Mycoparasitism: Some beneficial fungi can parasitize harmful fungi directly, a process called mycoparasitism. Trichoderma species, for instance, are known to attach themselves to the hyphae of pathogenic fungi, penetrate their cell walls, and consume their contents. This process weakens or kills the pathogen, protecting the host plant from disease.

Regulation of the Soil Microbial Community:

Fungi play a significant role in shaping the microbial community in the soil, fostering an environment that is less conducive to the growth of pathogens. Beneficial fungi help maintain a balanced soil ecosystem, where microbial diversity is high, and competition for resources is intense, limiting the opportunity for pathogens to dominate.

Suppressive soils: In some cases, soils that are rich in beneficial fungi can become "suppressive" to pathogens. This means that the microbial community, including fungi, bacteria, and other organisms, collectively creates conditions that inhibit the growth or activity of soil-borne pathogens. Fungal competition for organic matter and nutrients, along with the production of antimicrobial substances, plays a key role in creating suppressive soils.

Improved nutrient cycling and plant health: By enhancing nutrient availability, particularly nitrogen and phosphorus, mycorrhizal fungi contribute to overall plant vigor and health. Healthier plants are better equipped to resist pathogen attacks, as they have the energy and resources needed to maintain their immune defenses. Additionally, healthier root systems can more effectively ward off root-invading pathogens.

Fungi-Mediated Communication between Plants:

Another fascinating aspect of fungal interactions with plant immunity is their role in plant-to-plant communication. Through the mycorrhizal network, plants can signal each other about potential threats, such as the presence of pathogens or herbivores. This communication occurs when one plant in the network is attacked or infected and releases defense-related signaling molecules (e.g., jasmonic acid or volatile organic compounds) into the shared fungal hyphae. These signals are transmitted to neighboring plants, prompting them to activate their defense responses before they themselves are attacked.

Defense signaling across species: This signaling is not restricted to plants of the same species. Mycorrhizal networks can facilitate communication between different plant species, which allows entire plant communities to coordinate their defenses against pathogens. This shared immune response helps protect plants in the same ecosystem from widespread disease outbreaks.

Enhancing Systemic Resistance and Cross-Protection:

Beneficial fungi can also induce systemic resistance in plants, which extends beyond the immediate area of fungal colonization. This type of resistance, often referred to as induced systemic resistance (ISR), is triggered when plants detect beneficial microbes, including mycorrhizal fungi, and activate broad-spectrum immune responses that protect against a wide range of pathogens. Unlike the direct immune responses triggered by pathogens, ISR prepares the plant for future attacks by enhancing its overall immune system.

Cross-protection: Some fungi not only protect plants from fungal pathogens but also provide resistance against bacterial and viral pathogens. For example, plants colonized by arbuscular mycorrhizal fungi have been shown to exhibit enhanced resistance to bacterial infections like Ralstonia solanacearum and viral infections like tobacco mosaic virus (TMV). This cross-protection is thought to occur through the systemic activation of the plant's immune system, along with improved nutrient uptake and overall plant health.

Fungal Symbiosis and Pathogen Avoidance:

Fungal symbiosis can alter the physical and chemical environment around plant roots, making it more difficult for pathogens to establish infections. For example, mycorrhizal fungi can change the pH of the rhizosphere or release chemicals that inhibit pathogen activity. Moreover, fungal colonization of plant roots can occupy potential infection sites, effectively blocking pathogenic organisms from gaining access to the plant's tissues.

Fungi enhance plant immunity and pathogen resistance through a variety of mechanisms. Mycorrhizal fungi trigger immune responses and defense priming, while antagonistic fungi directly suppress pathogens through competition, mycoparasitism, and antimicrobial production. Additionally, fungi help regulate the soil microbial community, improve nutrient availability, and facilitate communication between plants, all of which contribute to a plant's overall ability to resist disease. The synergistic relationship between plants and beneficial fungi is crucial for maintaining plant health and protecting against a wide range of pathogens.

4.4. Conclusion

The complex and multifaceted interactions between fungi and plants are fundamental to the health and functionality of terrestrial ecosystems. This chapter has explored the various ways in which fungi contribute to plant growth, health, and resilience through their symbiotic, pathogenic, and endophytic relationships. **Mycorrhizal fungi** stand out as crucial partners in this dynamic, forming extensive networks that enhance nutrient and water uptake, particularly in nutrient-poor or stressful environments. By extending the root system and facilitating the acquisition of essential nutrients such as phosphorus and nitrogen, mycorrhizal fungi support robust plant growth and improve productivity. Additionally, their role in promoting plant resistance to drought, salinity, and heavy metals underscores their importance in sustaining plant health in diverse and challenging conditions.

Moreover, fungi's involvement extends beyond mere nutrient exchange; they are integral to plant immunity and stress resilience. The **systemic acquired resistance (SAR)** and **induced systemic resistance (ISR)** mechanisms activated by mycorrhizal fungi enable plants to better defend themselves against pathogens and pests. Similarly, **endophytic fungi** contribute to plant health by producing antimicrobial compounds and enhancing internal defense mechanisms, further safeguarding plants from biotic stressors. These interactions not only foster individual plant health but also promote ecosystem stability and biodiversity. Understanding these relationships highlights the intricate balance of mutual benefits and challenges that define the plant-fungi partnership. As research advances, it becomes increasingly evident that these fungal interactions are indispensable for sustainable agriculture, ecosystem management, and biodiversity conservation. Thus, appreciating and harnessing the power of these fungal-plant interactions is crucial for addressing future challenges in plant health and ecosystem resilience.

Chapter 5

Soil Fungi and Agricultural Practices

Soil fungi play a critical role in agricultural practices by significantly influencing soil health, fertility, and plant productivity. Mycorrhizal fungi, for example, form symbiotic relationships with crop roots, enhancing nutrient uptake, particularly phosphorus, which is often a limiting factor in agricultural soils. These fungi extend their hyphal networks into the soil, accessing nutrients beyond the reach of plant roots and improving the efficiency of nutrient utilization. This enhanced nutrient availability leads to improved crop growth, higher yields, and reduced need for chemical fertilizers, contributing to more sustainable farming practices. Furthermore, mycorrhizal fungi can enhance soil structure by forming aggregates through the secretion of glomalin, a protein that binds soil particles together. This improved soil structure enhances water infiltration, reduces erosion, and increases soil resilience against compaction, which is beneficial for overall crop health and productivity.

In addition to nutrient enhancement and soil structure improvement, soil fungi also play a role in disease suppression and stress mitigation. Certain soil fungi, such as beneficial *trichoderma species*, act as biological control agents against plant pathogens by competing for resources, producing antimicrobial compounds, and inducing plant defense mechanisms. This natural form of pest and disease management reduces the reliance on chemical pesticides, promoting a healthier and more environmentally friendly approach to agriculture. Moreover, soil fungi help plants cope with various stress conditions, including drought and salinity, by improving water and nutrient absorption. This stress resilience is crucial for maintaining crop yields in the face of changing climate conditions and ensuring food security. Integrating fungal management into agricultural practices, such as through the use of mycorrhizal inoculants and promoting fungal biodiversity in soil, supports sustainable farming by enhancing soil health, reducing chemical inputs, and improving overall crop performance.

Soil fungi are indispensable in agricultural practices due to their profound impact on soil fertility, plant health, and overall farm sustainability. Mycorrhizal fungi are particularly crucial as they establish mutualistic relationships with plant roots, enhancing nutrient uptake and improving soil structure. These fungi extend their hyphal networks into the soil, vastly increasing the surface area available for nutrient absorption. This is especially valuable for accessing immobile nutrients like phosphorus, which is often limited in agricultural soils. By improving nutrient availability, mycorrhizal fungi not only boost crop yields but also reduce the need for synthetic fertilizers, leading to cost savings and decreased environmental impact. Furthermore, the fungal hyphae contribute to soil aggregation by producing glomalin, a protein that binds soil particles together, enhancing soil structure, water retention, and resistance to erosion. This results in better soil aeration and reduced compaction, which promotes healthier root growth and improves overall crop resilience.

In addition to enhancing soil fertility and structure, soil fungi play a critical role in disease suppression and stress mitigation. Beneficial fungi, such as *Trichoderma* species, act as natural antagonists to harmful pathogens by competing for resources, producing antimicrobial compounds, and triggering plant defense mechanisms. This biological control reduces the reliance on chemical pesticides, supporting more sustainable and environmentally friendly farming practices. Soil fungi also help plants adapt to various stress conditions, such as drought and soil salinity, by improving water and nutrient uptake. This stress resilience is increasingly important in the context of climate change and variability, as it helps maintain stable crop yields and ensures food security. By integrating soil fungal management strategies, such as inoculating crops with beneficial fungi and fostering fungal diversity in soils, farmers can enhance soil health, promote sustainable agriculture, and improve crop productivity and resilience.

5.1. Importance of fungi in sustainable agriculture.

Fungi play a pivotal role in advancing sustainable agriculture by enhancing soil health, improving crop productivity, and reducing the reliance on chemical inputs. Mycorrhizal fungi, in particular, are integral to this process. These fungi form symbiotic relationships with plant roots, extending their hyphal networks into the soil. This relationship significantly boosts the plant's ability to absorb essential nutrients, particularly phosphorus, which is often limited in agricultural soils. The improved nutrient uptake not only enhances plant growth and crop yields but also reduces the need for synthetic fertilizers. This reduction in chemical fertilizer use mitigates the risk of soil and water pollution, which is crucial for maintaining environmental health and sustainability. Additionally, the hyphal networks of mycorrhizal fungi contribute to better soil structure by forming aggregates through the secretion of glomalin, a protein that binds soil particles. This improved soil structure enhances water retention, reduces erosion, and increases soil resilience to compaction, which benefits overall soil health and crop performance.

Beyond nutrient acquisition and soil structure, fungi are essential in promoting disease resistance and stress tolerance in crops. Beneficial fungi, such as **Trichoderma** and **Penicillium** species, act as natural biocontrol agents against soil-borne pathogens by competing for resources, producing antimicrobial compounds, and inducing systemic resistance in plants. This natural pest and disease management reduces the need for chemical pesticides, which can be harmful to ecosystems and human health. Furthermore, fungi enhance plant resilience to environmental stresses, such as drought, salinity, and extreme temperatures, by improving water and nutrient uptake and stimulating plant defense mechanisms. This stress tolerance is increasingly important as climate change leads to more frequent and severe weather events. By incorporating fungal management practices, such as using mycorrhizal inoculants and promoting fungal diversity in soils

85

sustainable agriculture can be advanced. These practices support healthier soils, reduce chemical inputs, and enhance crop productivity and resilience, contributing to more sustainable and environmentally friendly farming systems.

Fungi are central to the concept of sustainable agriculture due to their multifaceted roles in enhancing soil health, promoting plant growth, and reducing the environmental impact of farming practices. Their importance can be understood through several key contributions:

Nutrient Acquisition and Soil Fertility:

Mycorrhizal Fungi: These fungi form mutualistic relationships with plant roots, significantly improving the efficiency of nutrient uptake. Mycorrhizal fungi extend their hyphal networks beyond the reach of plant roots, accessing nutrients, especially phosphorus, that are otherwise immobile in the soil. This enhanced nutrient acquisition leads to improved plant growth and higher crop yields. Moreover, by reducing the need for synthetic phosphorus fertilizers, mycorrhizal fungi help decrease the risk of nutrient runoff, which can cause water pollution and eutrophication. The presence of mycorrhizal fungi also contributes to soil fertility by increasing the availability of other essential nutrients, such as nitrogen and trace minerals.

Soil Structure and Health:

Soil Aggregation: Mycorrhizal fungi and other soil fungi play a crucial role in soil aggregation by secreting substances like glomalin, which binds soil particles together. This enhances soil structure, improves water infiltration, reduces erosion, and increases soil resilience against compaction. Well-aggregated soil promotes healthy root growth, better aeration, and increased water retention, which are vital for maintaining soil health and supporting plant productivity.

Disease and Pest Management:

Biocontrol Agents: Certain fungi, such as **Trichoderma** and **Penicillium** species, act as natural biocontrol agents. They suppress soil-borne pathogens by outcompeting them for resources, producing antimicrobial compounds, and inducing plant defense mechanisms. This natural form of pest and disease management reduces the reliance on chemical pesticides, which can have harmful effects on the environment and human health. By fostering beneficial fungal populations, farmers can manage plant diseases more sustainably and reduce the need for chemical interventions.

Stress Tolerance and Resilience:

Enhanced Stress Tolerance: Soil fungi contribute to plant resilience under stressful conditions, such as drought, salinity, and extreme temperatures. Mycorrhizal fungi improve water and nutrient uptake, helping plants maintain growth and health during periods of environmental stress. This stress tolerance is increasingly important in the context of climate change, which brings more frequent and severe weather events. By enhancing plant resilience, fungi help ensure stable crop yields and food security.

Organic Matter Decomposition and Nutrient Recycling:

Decomposition: Soil fungi are key players in the decomposition of organic matter, breaking down plant residues and other organic materials into simpler forms. This process releases nutrients back into the soil, making them available for plant uptake. Effective decomposition by fungi contributes to nutrient cycling and the formation of humus, which enhances soil fertility and structure.

Promoting Biodiversity:

Ecological Balance: By supporting a diverse array of plant species through their various interactions, fungi contribute to overall ecosystem health and biodiversity. Diverse plant communities can better withstand pests, diseases, and environmental stresses, leading to more resilient agricultural systems. Fungal interactions help maintain ecological balance and promote sustainable agricultural practices by fostering a diverse and healthy soil microbiome.

Fungi play a vital role in sustainable agriculture by enhancing nutrient acquisition, improving soil health, managing pests and diseases, increasing stress tolerance, and contributing to nutrient recycling. Their ability to support plant growth while reducing the reliance on chemical inputs makes them indispensable partners in creating more sustainable and environmentally friendly farming systems.

5.2.Mycorrhizae and crop productivity.

Mycorrhizae, the symbiotic association between fungi and plant roots, significantly enhance crop productivity through their profound influence on nutrient acquisition and soil health. Mycorrhizal fungi extend their hyphal networks into the soil, far beyond the reach of plant roots, facilitating the uptake of essential nutrients, particularly phosphorus. Phosphorus, a critical nutrient for energy transfer, photosynthesis, and root development, is often present in low quantities in agricultural soils and can be challenging for plants to access. Mycorrhizal fungi mobilize phosphorus and other nutrients from soil particles and organic matter, increasing their availability to plants. This enhanced nutrient uptake leads to improved plant growth, more robust root systems, and higher crop yields. Additionally, by reducing the need for synthetic fertilizers, mycorrhizal associations contribute to more sustainable farming practices. The

87

reduced reliance on chemical inputs not only lowers production costs but also minimizes environmental impacts such as nutrient runoff and soil degradation.

Beyond nutrient acquisition, mycorrhizae play a crucial role in improving soil structure and enhancing plant resilience. The fungal hyphae secrete glomalin, a substance that binds soil particles together, leading to the formation of stable soil aggregates. This improved soil structure enhances water infiltration, reduces erosion, and decreases soil compaction, creating a more favorable environment for root growth. Better soil structure also improves water retention, which is particularly beneficial in drought-prone areas, as it helps maintain crop productivity under variable moisture conditions. Moreover, mycorrhizal fungi bolster plant health by enhancing disease resistance and stress tolerance. They suppress soil-borne pathogens through competition and antimicrobial activity, while also improving plant resilience to environmental stresses such as drought and salinity. By supporting more efficient nutrient and water use, promoting healthy soil, and enhancing plant defense mechanisms, mycorrhizae contribute to increased crop yields and overall agricultural sustainability.

Mycorrhizae, the symbiotic association between fungi and plant roots, play a crucial role in enhancing crop productivity through several mechanisms that improve nutrient uptake, soil health, and plant resilience.

Enhanced Nutrient Uptake:

Phosphorus Acquisition: Mycorrhizal fungi are particularly effective at facilitating the uptake of phosphorus, a vital nutrient that is often present in limited amounts in agricultural soils. The fungal hyphae extend far beyond the root zone of the plant, accessing phosphorus and other nutrients that are otherwise unavailable to the plant. This enhanced nutrient acquisition leads to improved plant growth and development. In phosphorus-deficient soils, mycorrhizal fungi can significantly boost crop yields by increasing the amount of available phosphorus, which is crucial for energy transfer, photosynthesis, and root development.

Other Nutrients: Besides phosphorus, mycorrhizal fungi also help in the uptake of other essential nutrients, including nitrogen, potassium, calcium, and trace minerals. They release enzymes that mobilize these nutrients from organic matter and soil minerals, making them more accessible to plants. This comprehensive nutrient support improves overall plant health and productivity.

Improved Soil Structure and Health:

Soil Aggregation: Mycorrhizal fungi contribute to the formation of stable soil aggregates through the secretion of substances like glomalin. This process enhances soil structure by binding soil particles together, which improves soil aeration, water infiltration, and root penetration. Well-structured soil supports healthy root growth and reduces the risk of soil erosion and compaction, creating a more favorable environment for crop growth.

Water Retention: The improved soil structure also enhances the soil's ability to retain water, reducing the frequency and intensity of irrigation required. This is particularly beneficial in areas prone to drought or where water resources are limited. By enhancing water availability, mycorrhizal fungi help maintain crop productivity under variable moisture conditions.

Disease Resistance and Stress Tolerance:

Disease Suppression: Mycorrhizal fungi can suppress soil-borne pathogens by competing for resources, producing antimicrobial compounds, and stimulating plant defense responses. This biological control reduces the incidence of plant diseases and minimizes the need for chemical pesticides. Healthier plants are more productive and require fewer chemical inputs, contributing to more sustainable farming practices.

Stress Resilience: Mycorrhizal fungi enhance plant resilience to various environmental stresses, including drought, salinity, and heavy metal contamination. By improving nutrient and water uptake, these fungi help plants maintain growth and productivity under stressful conditions. This increased stress tolerance is crucial for sustaining crop yields in the face of climate change and other environmental challenges.

Reduction in Chemical Inputs:

Fertilizer Efficiency: The enhanced nutrient uptake provided by mycorrhizal fungi reduces the need for synthetic fertilizers, which can be costly and environmentally harmful. By promoting efficient nutrient use, mycorrhizae help lower fertilizer applications, decrease nutrient runoff, and reduce the environmental impact of agricultural practices.

Improved Plant Health and Growth:

Root Development: Mycorrhizal fungi support more extensive and healthier root systems, which contribute to better plant growth and higher yields. The increased root surface area provided by fungal hyphae allows for more efficient water and nutrient absorption, supporting vigorous plant development.

Crop Yield: Numerous studies have demonstrated that crops associated with mycorrhizal fungi tend to exhibit higher yields compared to non-mycorrhizal crops. This increase in productivity is attributed to the combined benefits of improved nutrient uptake, better soil health, enhanced disease resistance, and greater stress tolerance.

Mycorrhizae play a pivotal role in boosting crop productivity by enhancing nutrient uptake, improving soil structure, supporting disease resistance, and increasing stress tolerance. Their contributions lead to healthier plants, reduced chemical inputs, and higher yields, making them an essential component of sustainable agricultural practices. By harnessing the benefits of mycorrhizal associations, farmers can achieve more efficient and environmentally friendly crop production.

5.3.Fungi as biofertilizers and biocontrol agents

Fungi play a crucial role as both biofertilizers and biocontrol agents, significantly contributing to sustainable agriculture by enhancing soil health and plant productivity while minimizing environmental impacts. As biofertilizers, fungi, particularly mycorrhizal fungi, form symbiotic relationships with plant roots, extending their hyphal networks into the soil to increase the surface area for nutrient absorption. This symbiosis is especially beneficial for acquiring immobile nutrients like phosphorus, which is essential for energy transfer, root development, and overall plant growth. Mycorrhizal fungi not only mobilize phosphorus from soil particles and organic matter, making it more accessible to plants, but also improve the uptake of other essential nutrients such as nitrogen, potassium, and trace minerals. Their presence enhances soil fertility by reducing the need for synthetic fertilizers, thus lowering agricultural costs and minimizing environmental issues like nutrient runoff and soil degradation. Additionally, mycorrhizal fungi contribute to soil health by improving soil structure through the secretion of glomalin, which binds soil particles into aggregates, enhancing water retention, aeration, and soil resilience. As biocontrol agents, certain fungi like *Trichoderma*, *Beauveria*, and *Coniothyrium* offer natural pest and disease management by competing with pathogens for resources, producing antimicrobial compounds, and inducing systemic resistance in plants. These beneficial fungi suppress soil-borne pathogens and pests, reducing the need for chemical pesticides and promoting environmentally friendly pest control methods. By supporting plant health and productivity through these dual roles, fungi help achieve more sustainable agricultural practices, ensuring increased crop yields while protecting the ecosystem.

Fungi as Biofertilizers

Fungi are increasingly recognized as effective biofertilizers due to their ability to enhance soil fertility and plant growth through symbiotic and associative relationships. One of the most prominent examples is mycorrhizal fungi, which form mutualistic associations with plant roots. These fungi extend their hyphal networks into the soil, increasing the surface area for nutrient absorption. This is particularly beneficial for acquiring immobile nutrients such as phosphorus, which is essential for energy transfer, root development, and overall plant growth. By mobilizing phosphorus from soil particles and organic matter, mycorrhizal fungi make this nutrient more accessible to plants, leading to improved growth and productivity.

Beyond phosphorus, mycorrhizal fungi also aid in the uptake of other essential nutrients, including nitrogen, potassium, calcium, and trace minerals. Their presence can enhance soil fertility by increasing the availability of these nutrients and reducing the need for synthetic fertilizers. This not only lowers agricultural costs but also minimizes the environmental impact of fertilizer use, such as nutrient runoff and soil degradation. Additionally, mycorrhizal fungi contribute to soil health by improving soil structure through the secretion of glomalin, a protein that binds soil particles together. This process enhances soil aggregation, water retention, and aeration, which are vital for sustaining plant growth and crop yields. Fungi play a pivotal role as biofertilizers by enhancing soil fertility and promoting plant growth through their symbiotic relationships with plant roots and their ability to influence soil chemistry and structure. This role is primarily exemplified by mycorrhizal fungi, but other fungal types also contribute significantly.

Mycorrhizal Fungi:

Symbiotic Relationships: Mycorrhizal fungi form mutualistic associations with plant roots, extending their hyphal networks into the soil. This relationship benefits plants by greatly increasing the effective surface area for nutrient absorption. The fungal hyphae penetrate the soil far beyond the root zone of the plant, accessing nutrients that are otherwise less available to plant roots. This is especially crucial for phosphorus, a key nutrient often present in limited amounts in soil. Mycorrhizal fungi release enzymes that solubilize phosphorus and other nutrients, making them more accessible to plants. This enhanced nutrient uptake supports better plant growth, root development, and overall crop productivity.

Nutrient Availability: Besides phosphorus, mycorrhizal fungi also facilitate the uptake of other essential nutrients such as nitrogen, potassium, calcium, and trace minerals. By mobilizing these nutrients from soil particles and organic matter, fungi improve soil fertility and plant nutrition. This reduces the need for synthetic fertilizers, which can be costly and environmentally damaging due to issues like nutrient runoff and soil acidification. By improving nutrient use efficiency, mycorrhizal fungi contribute to sustainable agricultural practices by lowering fertilizer requirements and mitigating environmental impacts.

Soil Structure Improvement:

Soil Aggregation: Mycorrhizal fungi contribute to the formation of stable soil aggregates through the production of glomalin, a protein that binds soil particles together. This process improves soil structure, leading to better soil aeration, water infiltration, and reduced erosion. Well-structured soil enhances root growth and allows for more efficient water and nutrient uptake, further supporting plant health and productivity. The improved soil structure also helps in maintaining optimal soil moisture levels, which is beneficial in both drought and heavy rainfall conditions.

Organic Matter Decomposition:

Decomposition: Fungi play a critical role in decomposing organic matter, such as plant residues and other organic materials. This decomposition process releases nutrients back into the soil, making them available for plant uptake. Decomposition by fungi contributes to nutrient cycling and the formation of humus, which enhances soil fertility and structure. By accelerating the breakdown of organic matter, fungi help maintain nutrient availability and support healthy plant growth.

Disease Suppression:

Disease Management: While primarily a function of biocontrol, fungi also indirectly support their role as biofertilizers by improving plant health and reducing the incidence of soil-borne diseases. Healthy plants with improved nutrient uptake are better equipped to resist diseases. Additionally, some fungi produce antimicrobial compounds that suppress pathogens, further contributing to plant health and reducing the need for chemical fungicides.

Stress Tolerance:

Environmental Stress: Mycorrhizal fungi enhance plant resilience to environmental stresses such as drought, salinity, and heavy metal contamination. By improving water and nutrient uptake, these fungi help plants maintain growth and health under adverse conditions. This increased stress tolerance is particularly valuable in the context of climate change, which presents more frequent and severe weather events.

Fungi as biofertilizers play a multifaceted role in sustainable agriculture by enhancing nutrient availability, improving soil structure, supporting organic matter decomposition, and contributing to plant health and stress tolerance. Their ability to reduce the reliance on synthetic fertilizers and mitigate environmental impacts makes them a valuable component of integrated and sustainable farming practices.

Fungi as Biocontrol Agents

Fungi are also valuable biocontrol agents in managing plant diseases and pests. Certain fungal species, such as *Trichoderma, Beauveria, and Coniothyrium*, are employed in integrated pest management (IPM) systems to suppress soil-borne pathogens and harmful insects. These fungi can outcompete pathogens for resources, produce antifungal or antibacterial compounds, and induce plant defense mechanisms. For example, Trichoderma species are known for their ability to colonize plant roots and produce enzymes that degrade the cell walls of pathogens, thereby preventing their growth and spread. Some fungi can enhance plant immunity by triggering systemic resistance responses. This involves the activation of the plant's defense systems, leading to increased resistance against a broad range of pathogens and pests. This natural form of pest and disease control reduces the need for chemical pesticides, which can be harmful to ecosystems and human health. By promoting the use of beneficial fungi as biocontrol agents, farmers can achieve more sustainable and environmentally friendly pest management solutions. These fungal biocontrol agents help maintain plant health and productivity while minimizing the reliance on chemical inputs, thus contributing to more resilient and sustainable agricultural systems. Fungi play a crucial role as biocontrol agents in managing plant diseases and pests, providing an environmentally friendly alternative to chemical pesticides. They contribute to integrated pest management (IPM) systems through various mechanisms, including direct antagonism of pathogens, competition for resources, and induction of plant defense responses.

Direct Antagonism:

Pathogen Suppression: Certain fungi, such as *Trichoderma* species, act directly against plant pathogens through various mechanisms. These fungi produce a range of antimicrobial compounds, including antibiotics and enzymes, which degrade the cell walls of pathogens, thereby inhibiting their growth and reducing their ability to infect plants. For instance, *Trichoderma harzianum* secretes chitinases and glucanases that break down the cell walls of fungi like *Fusarium* and *Rhizoctonia*, common soil-borne pathogens. This direct antagonism helps prevent the establishment and spread of harmful pathogens, leading to healthier plants and reduced disease incidence.

Competition for Resources:

Nutrient and Space Competition: Biocontrol fungi can outcompete plant pathogens for essential nutrients and space in the soil or on plant surfaces. By colonizing the root zone or plant surfaces more effectively than pathogens, these fungi limit the resources available to harmful microorganisms. For example, fungi like *Bacillus subtilis* and *Pseudomonas fluorescens* establish themselves on plant roots and create a competitive environment that restricts pathogen colonization. This competitive exclusion reduces the pathogen load and lowers the risk of disease outbreaks.

Induction of Plant Defense Responses:

Systemic Resistance: Some biocontrol fungi can induce systemic resistance in plants, enhancing their ability to fend off a broad spectrum of pathogens. This process involves the activation of plant defense mechanisms, such as the production of defensive proteins, enzymes, and secondary metabolites. For instance, *Beauveria bassiana*, an entomopathogenic fungus, can trigger systemic acquired resistance (SAR) in plants, which helps them resist not only the pathogen it targets but also other unrelated pathogens. This induced resistance enhances the plant's overall disease resilience and reduces the need for chemical fungicides.

Biological Control of Insect Pests:

Insect Pathogens: Certain fungi, such as *Beauveria bassiana* and *Metarhizium anisopliae*, are effective biocontrol agents against insect pests. These fungi infect and kill insects by penetrating their exoskeletons and proliferating inside their bodies. The spores of these fungi attach to the insect's cuticle, germinate, and grow, eventually leading to the insect's death. This form of biological control is particularly useful in managing pests like aphids, beetles, and caterpillars, providing an alternative to chemical insecticides and helping maintain ecological balance in agricultural systems.

Soil Health Improvement:

Soil Microbial Community: By promoting a diverse and healthy soil microbial community, biocontrol fungi contribute to overall soil health, which indirectly supports plant health. Healthy soils with a balanced microbial community are better able to suppress pathogens and support plant growth. The presence of beneficial fungi helps maintain this balance by outcompeting harmful microbes and enhancing soil structure and nutrient availability.

Sustainable Agricultural Practices:

Reduced Chemical Use: The use of fungal biocontrol agents reduces reliance on synthetic chemicals, which can have adverse environmental and health effects. By incorporating biocontrol fungi into pest and disease management strategies, farmers can decrease the use of chemical pesticides and fungicides, leading to more sustainable agricultural practices. This approach not only minimizes the risk of chemical resistance in pathogens but also reduces environmental pollution and promotes ecological balance.

Fungi serve as effective biocontrol agents by directly antagonizing pathogens, competing for resources, inducing plant defense mechanisms, and controlling insect pests. Their role in reducing chemical inputs and promoting soil health underscores their importance in sustainable agriculture. By leveraging fungal biocontrol agents, farmers can enhance crop protection, support plant health, and contribute to environmentally friendly farming practices. Fungi serve dual roles as biofertilizers and biocontrol agents, enhancing soil fertility and plant health while promoting sustainable agricultural practices. Their ability to improve nutrient uptake, soil structure, and disease resistance makes them valuable tools in modern farming, supporting both increased crop productivity and environmental stewardship.

5.4.Conclusion

Soil fungi, particularly mycorrhizal fungi, significantly contribute to improved nutrient acquisition, soil structure, and overall plant health. Mycorrhizal associations enhance the uptake of vital nutrients such as phosphorus, nitrogen, and other minerals, which are often limited in agricultural soils. By extending their hyphal networks into the soil, these fungi mobilize nutrients that are otherwise inaccessible to plant roots, thus boosting plant growth and crop yields. Furthermore, the production of glomalin by mycorrhizal fungi improves soil aggregation, leading to better water retention, reduced erosion, and enhanced soil resilience. This ability to improve soil health and reduce dependency on synthetic fertilizers is pivotal in promoting sustainable agricultural practices, as it reduces environmental impacts associated with chemical inputs and supports more efficient nutrient use.

In addition, soil fungi serve as valuable biocontrol agents, offering natural solutions for managing plant diseases and pests. Through mechanisms such as direct antagonism of pathogens, competition for resources, and induction of systemic resistance, fungi can suppress harmful microorganisms and pests, thereby reducing the need for chemical pesticides. This not only helps in controlling pest and disease outbreaks but also fosters a more balanced and resilient agricultural ecosystem. By integrating fungal biocontrol agents into pest management strategies, farmers can achieve greater crop protection while minimizing environmental and health risks associated with chemical treatments. The use of fungi in agriculture not only contributes to increased crop yields and improved soil conditions but also aligns with ecological principles, fostering more sustainable and environmentally responsible farming practices

Chapter 6

Fungi and Soil Health in Natural Ecosystems

Fungi are integral to maintaining soil health in natural ecosystems through their diverse roles in nutrient cycling, soil structure formation, and organic matter decomposition. As decomposers, fungi break down complex organic materials, such as plant residues, dead animals, and leaf litter, into simpler compounds. This process is essential for recycling nutrients, making them available for uptake by plants. Fungi decompose organic matter into humus, which enhances soil fertility and structure. The decomposition process also contributes to the formation of soil aggregates, which improve soil aeration, water infiltration, and stability. By facilitating the breakdown of organic matter and the release of essential nutrients, fungi support the productivity and resilience of natural ecosystems. Their ability to convert organic material into forms accessible to plants is fundamental to the health and sustainability of forest floors, grasslands, and other natural habitats.

In addition to nutrient cycling, fungi play a crucial role in enhancing soil structure and supporting soil biodiversity. Mycorrhizal fungi, in particular, form symbiotic relationships with plant roots, extending their hyphal networks into the soil. This not only improves the uptake of nutrients such as phosphorus and nitrogen but also contributes to soil aggregation through the secretion of glomalin, a substance that binds soil particles together. This improved soil structure enhances water retention, reduces erosion, and promotes a healthier root environment. Furthermore, fungi support a diverse soil microbiome by interacting with other microorganisms, creating a balanced ecosystem that aids in disease suppression and enhances soil fertility. Through these roles, fungi help maintain the intricate balance of natural ecosystems, ensuring that soils remain productive, resilient, and capable of supporting a wide range of plant and animal life. Their contributions are essential for sustaining the health and functionality of natural environments, making them a key component in ecosystem management and conservation efforts.

Fungi are vital to soil health in natural ecosystems due to their fundamental roles in nutrient cycling, soil structure formation, and the maintenance of ecological balance. As primary decomposers, fungi break down complex organic materials like plant residues and dead organisms into simpler compounds, recycling essential nutrients and converting them into forms accessible to plants. This decomposition process not only replenishes soil nutrients but also contributes to the formation of humus, which improves soil fertility and structure. Mycorrhizal fungi, which form symbiotic relationships with plant roots, enhance nutrient uptake by extending their hyphal networks into the soil, thus increasing the availability of critical nutrients such as phosphorus and nitrogen. Their secretion of glomalin binds soil particles together, leading to improved soil aggregation, better water retention, and reduced erosion. Additionally, fungi support soil biodiversity by interacting with other microorganisms, fostering a balanced and resilient microbial community that aids in disease suppression and overall soil health. Through these processes, fungi play an indispensable role in sustaining the productivity and stability of natural ecosystems, ensuring that soils remain fertile, resilient, and capable of supporting diverse plant and animal life.

6.1.How fungi maintain soil health in forests, grasslands, and other ecosystems.

In forest ecosystems, fungi are pivotal to soil health through their roles in decomposition, nutrient cycling, and soil structure enhancement. Forest soils are often rich in organic matter from fallen leaves, branches, and dead trees. Fungi, particularly decomposer fungi, break down this organic matter into simpler compounds, releasing essential nutrients like nitrogen, phosphorus, and potassium back into the soil. This nutrient recycling is crucial for sustaining the fertility of forest soils, which often have a limited natural nutrient supply. Mycorrhizal fungi, which form symbiotic relationships with tree roots, extend their hyphal networks into the soil, dramatically increasing the surface area for nutrient absorption. These fungi enhance the uptake of vital nutrients, especially phosphorus, which is frequently bound to soil particles and thus less accessible to plants. Mycorrhizal associations also contribute to soil structure by producing glomalin, a protein that helps bind soil particles into aggregates. This improves soil aeration, water retention, and reduces erosion, all of which support healthy forest ecosystems and biodiversity.

In grasslands and various other ecosystems, fungi play a similar role in maintaining soil health through their decomposition and nutrient cycling activities. Grasslands are characterized by a high turnover of plant material, including roots and above-ground biomass. Fungi decompose this organic material, converting it into humus, which enriches the soil with nutrients and improves its structure. This decomposition process also enhances the soil's water-holding capacity and resilience against erosion. Mycorrhizal fungi associated with grassland plants support nutrient uptake, particularly phosphorus, and help plants adapt to environmental stresses such as drought and nutrient deficiencies. Fungal networks also interact with other soil microorganisms, creating a balanced and robust microbial community that supports plant health and suppresses soil-borne pathogens. By contributing to nutrient availability, soil structure, and microbial balance, fungi ensure that grasslands and other ecosystems remain productive and resilient, supporting a diverse range of plant and animal life while maintaining overall soil health.

Fungi's Role in Maintaining Soil Health in Forest Ecosystems

In forest ecosystems, fungi are indispensable for maintaining soil health through several critical functions related to decomposition, nutrient cycling, soil structure, and ecological interaction

Decomposition and Nutrient Cycling:

Organic Matter Breakdown: Forest floors accumulate significant amounts of organic matter from fallen leaves, dead trees, branches, and other plant residues. Fungi, particularly decomposer fungi, play a central role in breaking down this organic material. They secrete enzymes that degrade complex organic compounds such as lignin and cellulose, transforming them into simpler molecules. This process not only releases essential nutrients like nitrogen, phosphorus, and potassium back into the soil but also converts organic matter into humus, a stable form of organic matter that enriches soil fertility. By facilitating nutrient cycling, fungi help maintain a continuous supply of vital nutrients for plant growth, which is crucial in forest ecosystems where nutrient availability can be limited.

Mycorrhizal Associations:

Symbiotic Relationships: Mycorrhizal fungi form mutualistic associations with the roots of most forest plants, including trees and understory vegetation. These fungi extend their hyphal networks into the soil, greatly increasing the effective surface area for nutrient absorption. This is particularly important for the uptake of phosphorus, which is often present in forms that are poorly available to plants. Mycorrhizal fungi solubilize phosphorus and other nutrients, making them accessible to plant roots. In return, plants provide the fungi with carbohydrates and other organic compounds necessary for their growth. This symbiosis enhances nutrient uptake, promotes vigorous plant growth, and contributes to forest productivity.

Soil Structure and Stability:

Soil Aggregation: Fungi contribute to soil structure through the production of glomalin, a glycoprotein produced by mycorrhizal fungi. Glomalin binds soil particles together, forming stable soil aggregates. This aggregation improves soil porosity, enhances water infiltration and retention, and reduces erosion. Stable soil structure is essential for preventing soil compaction, which can hinder root growth and reduce plant productivity. By promoting soil aggregation, fungi help maintain a healthy and functional soil environment that supports the diverse plant life found in forests.

Ecological Interactions:

Microbial Community Support: Fungi interact with other soil microorganisms, including bacteria and other fungi, to create a balanced microbial community. This microbial network supports various ecological functions, including disease suppression and soil health maintenance. For example, mycorrhizal fungi can suppress certain soil-borne pathogens by outcompeting them for resources or by producing antimicrobial compounds. A diverse and balanced microbial community enhances soil resilience, supports plant health, and contributes to overall forest ecosystem stability.

Soil Carbon Sequestration:

Carbon Storage: Fungi contribute to soil carbon sequestration through their decomposition activities. As they break down organic matter, they help incorporate carbon into the soil in the form of stable organic compounds. This process not only improves soil fertility but also plays a role in mitigating climate change by capturing carbon dioxide from the atmosphere and storing it in the soil.

Fungi play a multifaceted role in maintaining soil health in forest ecosystems by facilitating decomposition, enhancing nutrient availability through mycorrhizal associations, improving soil structure, supporting microbial diversity, and contributing to soil carbon storage. Their activities ensure that forest soils remain fertile, resilient, and capable of supporting diverse plant and animal life, thereby sustaining the overall health and functionality of forest ecosystems.

Fungi's Role in Maintaining Soil Health in Grasslands

Fungi play a crucial role in maintaining soil health in grassland ecosystems through their involvement in decomposition, nutrient cycling, soil structure, and symbiotic relationships with plants. These functions contribute to the overall productivity and resilience of grasslands, which are characterized by their dense cover of grasses and other herbaceous plants.

Decomposition and Nutrient Cycling:

Organic Matter Breakdown: Grasslands produce substantial amounts of organic matter through plant residues such as dead leaves, roots, and stems. Fungi are key decomposers that break down this organic material into simpler compounds. They secrete a variety of enzymes, such as cellulases and ligninases, that degrade complex substances like cellulose and lignin found in plant tissues. This decomposition process converts organic matter into humus, a stable form of organic material that enriches the soil with nutrients. By recycling nutrients such as nitrogen, phosphorus, and potassium, fungi ensure that these essential elements are available for uptake by plants, thereby supporting healthy grassland growth and productivity.

Mycorrhizal Associations:

Symbiotic Relationships: In grasslands, mycorrhizal fungi form symbiotic relationships with the roots of grasses and other herbaceous plants. These fungi extend their hyphal networks into the soil, increasing the surface area for nutrient absorption. Mycorrhizae are particularly important for enhancing the uptake of phosphorus, which is often bound in forms that are less accessible to plant roots. By solubilizing and mobilizing phosphorus, mycorrhizal fungi make it available to plants, promoting vigorous growth and resilience. In return, plants supply the fungi with carbohydrates and other organic compounds necessary for their growth. This mutualistic interaction contributes to the overall health and productivity of grassland ecosystems.

Soil Structure and Stability:

Soil Aggregation: Fungi contribute to soil structure through their ability to produce glomalin, a glycoprotein secreted by mycorrhizal fungi. Glomalin binds soil particles together, forming stable soil aggregates. These aggregates improve soil porosity, water infiltration, and aeration, which are critical for plant root health and soil stability. Enhanced soil structure reduces erosion, prevents soil compaction, and supports a healthy root environment. By promoting soil aggregation, fungi play a vital role in maintaining soil health and functionality in grassland ecosystems.

Support for Soil Biodiversity:

Microbial Interactions: Fungi interact with other soil microorganisms, including bacteria and other fungi, to create a balanced and diverse microbial community. This microbial network supports various ecological functions, such as nutrient cycling, disease suppression, and soil health maintenance. For example, mycorrhizal fungi can help suppress certain soil-borne pathogens by competing for resources or by producing antimicrobial compounds. A diverse and balanced microbial community enhances the resilience of grassland soils, supporting plant health and contributing to the overall stability of the ecosystem.

Stress Tolerance:

Drought and Nutrient Deficiencies: Mycorrhizal fungi help grassland plants tolerate environmental stresses such as drought and nutrient deficiencies. By extending their hyphal networks into the soil, these fungi improve water and nutrient uptake, which helps plants survive adverse conditions. This enhanced stress tolerance contributes to the resilience of grassland ecosystems, allowing them to withstand and recover from environmental challenges.

Fungi are essential for maintaining soil health in grasslands through their roles in decomposition, nutrient cycling, mycorrhizal symbiosis, soil structure improvement, and support for microbial diversity. Their activities ensure that grassland soils remain fertile, resilient, and capable of supporting diverse plant and animal life, thereby sustaining the overall health and functionality of these vital ecosystems.

Fungi's Role in Maintaining Soil Health in Various Ecosystems

Fungi contribute significantly to soil health across a range of ecosystems beyond forests and grasslands, including wetlands, deserts, and tundras. Their roles in decomposition, nutrient cycling, soil structure formation, and ecological interactions are crucial for sustaining the productivity and resilience of these diverse environments.

Wetlands:

Decomposition and Nutrient Cycling: In wetlands, fungi are essential decomposers that break down organic matter such as plant detritus, fallen leaves, and animal remains. Wetland soils often have high organic content due to the accumulation of plant material in waterlogged conditions. Fungi decompose this organic material, releasing nutrients such as nitrogen and phosphorus back into the soil. This nutrient cycling supports plant growth and maintains the fertility of wetland soils. Additionally, the anaerobic conditions in wetlands can slow down decomposition, but fungi adapted to these environments can still effectively break down organic matter, ensuring that nutrient cycles remain functional.

Soil Structure and Water Regulation: Fungi also contribute to soil structure in wetlands by producing substances that help bind soil particles together, forming aggregates that improve soil porosity and reduce compaction. This aggregation enhances water retention and reduces erosion, which is critical for maintaining the stability of wetland soils. Fungi play a role in regulating water levels and reducing soil erosion, which is important for the stability and health of wetland ecosystems.

Deserts:

Nutrient Acquisition and Soil Formation: In arid desert environments, fungi are important for nutrient acquisition and soil formation. The harsh conditions and low organic matter content make nutrient availability limited, but fungi help to break down sparse organic material, such as plant roots and decomposing animals, into simpler compounds. Mycorrhizal fungi form symbiotic relationships with desert plants, enhancing their ability to absorb nutrients and water from the soil. These fungi extend their hyphal networks into the soil, increasing nutrient uptake and helping plants survive in nutrient-poor conditions.

Soil Stabilization: Fungi contribute to soil stabilization in deserts by producing mycelial networks that bind soil particles together. This mycelial mat helps to prevent soil erosion by stabilizing loose, sandy soils and reducing wind and water erosion. In addition, the formation of soil aggregates through fungal activity improves water retention, which is crucial for sustaining plant life in arid environments.

Tundras:

Decomposition and Carbon Storage: In tundra ecosystems, fungi are vital for the decomposition of organic matter in cold, often waterlogged conditions. The decomposition process in tundra soils is slower due to low temperatures, but fungi are adapted to these conditions and continue to break down plant residues and other organic materials. This decomposition releases nutrients and contributes to the formation of peaty soils, which store large amounts of carbon. By facilitating the breakdown of organic matter, fungi help regulate the carbon cycle and contribute to the global carbon balance.

Symbiotic Relationships: Mycorrhizal fungi in tundras form symbiotic relationships with the roots of tundra plants, including mosses and lichens. These fungi improve nutrient uptake, particularly phosphorus, which is often limited in tundra soils. The symbiosis enhances plant growth and survival in nutrient-poor conditions, contributing to the overall health and productivity of tundra ecosystems.

Riparian Zones:

Nutrient Filtering and Soil Health: In riparian zones, which are the interface between land and water bodies, fungi play a crucial role in filtering nutrients and organic matter from runoff before it enters aquatic systems. This process helps prevent nutrient pollution in water bodies and maintains soil health by breaking down organic material and recycling nutrients. Fungi contribute to the stability of riparian soils by promoting soil aggregation and reducing erosion, which helps protect water quality and supports diverse plant and animal life.

Fungi maintain soil health in various ecosystems through their roles in decomposition, nutrient cycling, soil structure formation, and symbiotic relationships with plants. Their activities help sustain the fertility, stability, and resilience of soils across diverse environments, from wetlands and deserts to tundras and riparian zones, ensuring the overall health and functionality of these vital ecosystems.

6.2. Fungal adaptation to extreme environments (deserts, tundras).

Fungi exhibit remarkable adaptations that enable them to thrive in extreme environments where other organisms might struggle to survive. In arid environments such as deserts, fungi have evolved to cope with extreme dryness and temperature fluctuations. They possess specialized adaptations to manage water scarcity, including the ability to enter a dormant state during prolonged dry periods. This dormancy allows them to conserve energy and moisture until conditions become favorable again. Desert fungi often have highly efficient mechanisms for water absorption, such as extensive mycelial networks that can access moisture from deeper soil layers. They also produce protective structures like spores that can withstand desiccation and extreme temperatures. These adaptations enable fungi to continue their metabolic activities and reproductive processes even under harsh conditions, contributing to their survival and ecological roles in arid landscapes.

In cold environments, such as tundras and high-altitude regions, fungi have developed strategies to endure freezing temperatures and limited nutrient availability. Psychrophilic fungi, which are adapted to cold environments, produce enzymes that function optimally at low temperatures and help prevent ice formation within their cells. These enzymes are crucial for maintaining metabolic processes and facilitating decomposition even in frigid conditions. Additionally, fungi in cold environments often form symbiotic relationships with plants, such as mycorrhizal associations, to enhance nutrient uptake and support plant growth. In these symbiotic relationships, fungi provide plants with essential nutrients while benefiting from the organic compounds supplied by their plant partners. These adaptations enable fungi to maintain their ecological functions and contribute to the stability and productivity of cold ecosystems. Overall, fungal adaptations to extreme environments highlight their evolutionary versatility and their crucial roles in diverse and challenging ecological niches.

Fungal Adaptation to Extreme Environments: Deserts

Desert environments present extreme conditions, including intense heat, low humidity, and limited nutrient availability. Fungi that thrive in these arid regions exhibit a range of specialized adaptations to survive and maintain ecological functions in such harsh conditions.

Water Conservation and Absorption:

Dormancy and Resilience: Desert fungi have adapted to cope with prolonged periods of drought by entering a state of dormancy. During these dry spells, they conserve their metabolic energy and moisture, effectively halting their growth and reproductive activities until favorable conditions return. When moisture becomes available, these fungi can quickly reactivate their metabolic processes and resume growth. This adaptation is crucial for surviving the infrequent but intense rainfall typical of desert environments.

Efficient Water Use: Fungi in deserts often possess highly efficient mechanisms for water uptake. They develop extensive mycelial networks that can access moisture from deeper soil layers, beyond the reach of plant roots. This ability to explore a larger volume of soil helps fungi find and utilize the limited water resources available in arid conditions. Additionally, desert fungi may have structures like specialized hyphal tips that enhance their ability to absorb water from the soil.

Protection Against Desiccation and Temperature Extremes:

Desiccation Tolerance: To withstand the low humidity and high temperatures characteristic of desert environments, fungi produce highly resistant spores or resting structures. These spores are equipped with protective coatings that prevent water loss and shield the fungal cells from extreme heat. The ability of these spores to remain viable for extended periods of time without moisture ensures that fungi can survive until more favorable conditions occur.

Thermal Tolerance: Desert fungi have evolved to endure high temperatures by producing heat-stable enzymes and proteins that remain functional even in extreme heat. These enzymes facilitate essential metabolic processes and enable the fungi to continue their growth and reproduction during periods of intense heat. Additionally, some desert fungi have adapted to regulate their internal temperature and prevent damage from overheating, further enhancing their survival in extreme conditions.

Nutrient Acquisition and Soil Interaction:

Resource Utilization: The nutrient-poor conditions in deserts require fungi to be highly efficient in acquiring and utilizing available nutrients. They often form mutualistic relationships with desert plants, such as mycorrhizal associations, to enhance nutrient uptake. Mycorrhizal fungi extend their hyphal networks into the soil, increasing the surface area for nutrient absorption and facilitating the uptake of essential minerals like phosphorus and nitrogen that are limited in desert soils.

Soil Aggregation: Desert fungi contribute to soil health by promoting soil aggregation through the production of mycelial networks and binding agents. These fungal structures help stabilize loose, sandy soils, reducing erosion and improving soil structure. The enhanced soil aggregation also improves water retention, which is crucial for the survival of both fungi and desert plants.

Fungi in desert environments exhibit a range of adaptations that enable them to survive and thrive in extreme conditions. These adaptations include mechanisms for water conservation, protection against desiccation and high temperatures, efficient nutrient acquisition, and contributions to soil stability. Through these specialized strategies, desert fungi play vital roles in maintaining the ecological balance and productivity of arid ecosystems.

Fungal Adaptation to Extreme Environments: Tundras

Tundra ecosystems are characterized by extreme cold temperatures, short growing seasons, and often nutrient-poor soils. Fungi in tundras have developed a variety of adaptations to survive and thrive in these challenging conditions. Their strategies encompass metabolic adaptations, symbiotic relationships, and structural modifications that ensure their continued function and ecological role.

Adaptations to Cold Temperatures:

Psychrophilic Enzymes: Fungi in tundra environments, known as psychrophiles, have adapted to low temperatures by producing enzymes that are active and efficient at cold temperatures. These cold-adapted enzymes facilitate essential biochemical processes such as nutrient breakdown and energy production even when the external temperatures are extremely low. The ability of these enzymes to remain functional at low temperatures is crucial for maintaining fungal activity and decomposition processes in the short growing season of the tundra.

Cryoprotectants: To prevent cellular damage from freezing temperatures, tundra fungi produce cryoprotectants—substances that protect their cells from ice formation. These cryoprotectants include compounds such as trehalose and glycerol, which help to stabilize cell membranes and proteins, thereby preventing ice crystal formation that could disrupt cellular function. This adaptation allows fungi to remain viable and continue their metabolic activities despite the freezing conditions.

Nutrient Acquisition and Symbiotic Relationships:

Decomposition in Cold Soils: In tundras, the decomposition of organic matter is slower due to the cold temperatures and limited microbial activity. However, fungi are well adapted to these conditions and play a crucial role in breaking down plant litter and other organic material. By decomposing this material, fungi release essential nutrients into the soil, which is crucial for the limited plant life that exists in tundra ecosystems.

Mycorrhizal Associations: Tundra fungi often form mutualistic mycorrhizal relationships with plants, such as mosses, lichens, and other low-growing vegetation. These fungi extend their hyphal networks into the soil, increasing the effective surface area for nutrient and water uptake. This relationship is particularly important in tundras where nutrient availability is limited. Mycorrhizal fungi enhance the uptake of essential nutrients like phosphorus, which is often bound in forms that are less accessible to plant roots. In return, the plants provide the fungi with carbohydrates and organic compounds necessary for their growth. This symbiosis is critical for the survival and productivity of tundra plants.

Soil Formation and Carbon Storage:

Peat Formation: Tundra soils are often characterized by the accumulation of peat, a type of organic material that forms from the slow decomposition of plant matter in waterlogged conditions. Fungi contribute to the formation of peat by decomposing plant residues and other organic matter. The slow decomposition process in cold, waterlogged soils results in the accumulation of partially decomposed organic material, which eventually forms peat. This process contributes to carbon sequestration, as the organic carbon remains stored in the soil rather than being released into the atmosphere.

Soil Stabilization: Fungi play a role in stabilizing tundra soils through the formation of mycelial networks that bind soil particles together. These fungal networks help to prevent soil erosion and maintain soil structure, which is important in the fragile tundra ecosystem. By stabilizing the soil, fungi contribute to the overall health and functionality of tundra environments, supporting plant life and maintaining the ecological balance.

Fungi in tundra environments have evolved specialized adaptations to cope with extreme cold, slow decomposition rates, and nutrient limitations. Their adaptations include producing cold-active enzymes, using cryoprotectants to prevent freezing damage, forming mycorrhizal associations with plants, and contributing to peat formation and soil stabilization. These strategies enable fungi to play a crucial role in nutrient cycling, soil formation, and the maintenance of ecological balance in the harsh conditions of tundra ecosystems.

6.3.Conservation of fungal biodiversity.

Conserving fungal biodiversity is crucial for maintaining ecosystem health, stability, and functionality. Fungi play essential roles in nutrient cycling, soil formation, plant health, and ecological interactions, making their preservation vital for the overall well-being of ecosystems. The conservation of fungal biodiversity involves addressing several key aspects:

Understanding and Documenting Fungal Diversity:

Surveying and Inventorying: One of the first steps in fungal conservation is to document and inventory fungal species and their distributions. This involves conducting field surveys and collecting samples from various habitats to identify and catalog fungal species. Comprehensive fungal inventories help to understand the diversity of fungi within different ecosystems, identify species of conservation concern, and track changes in fungal populations over time. Researchers use molecular techniques, such as DNA sequencing, to accurately identify fungi and uncover previously unknown species.

Ecological Role Assessment: Understanding the ecological roles of fungi is essential for their conservation. Researchers study how fungi contribute to nutrient cycling, soil health, plant interactions, and other ecosystem functions. This knowledge helps prioritize conservation efforts for fungi that play critical roles in maintaining ecosystem processes and functions. Assessing the ecological importance of fungi also informs management strategies and conservation planning.

Threats to Fungal Biodiversity:

Habitat Destruction: Habitat loss due to deforestation, urbanization, agriculture, and other land-use changes poses a significant threat to fungal biodiversity. Many fungi have specialized habitat requirements and are sensitive to changes in their environment. Conservation efforts must address habitat preservation and restoration to protect fungal species and their ecosystems. Establishing protected areas and implementing sustainable land management practices can help mitigate habitat destruction and preserve critical fungal habitats.

Climate Change: Climate change affects fungal biodiversity by altering temperature, precipitation patterns, and seasonal dynamics. Changes in climate can impact fungal growth, reproduction, and distribution, as well as their interactions with other organisms. Monitoring and modeling the effects of climate change on fungi can help predict potential impacts and guide conservation strategies. Adaptive management approaches, such as adjusting conservation practices in response to changing climate conditions, are important for addressing climate-related threats to fungal biodiversity.

Pollution and Contamination: Pollution from chemicals, heavy metals, and other contaminants can negatively impact fungal communities and their functions. Pollution can alter soil chemistry, reduce nutrient availability, and harm fungal growth and reproduction. Conservation efforts should include measures to reduce pollution and manage contaminants to protect fungal species and maintain ecosystem health. Implementing best practices for waste management, reducing pesticide use, and monitoring soil and water quality are essential for mitigating pollution-related threats to fungi.

Conservation Strategies and Practices:

Protected Areas and Reserves: Establishing protected areas and reserves is an effective strategy for conserving fungal biodiversity. These areas provide refuge for fungal species and their habitats, reducing the impact of human activities and habitat loss. Protected areas can also serve as sites for research and monitoring, helping to track changes in fungal populations and assess the effectiveness of conservation measures.

Restoration and Rehabilitation: Habitat restoration and rehabilitation projects aim to restore degraded ecosystems and improve conditions for fungal communities. Restoration efforts may involve reforestation, soil improvement, and the reintroduction of native plant species to create suitable habitats for fungi. Effective restoration projects should consider the specific needs of fungal species and their ecological roles.

Public Awareness and Education: Raising public awareness about the importance of fungal biodiversity and its conservation is crucial for garnering support and promoting sustainable practices. Educational programs, outreach activities, and citizen science initiatives can help increase knowledge about fungi and their roles in ecosystems. Engaging the public in conservation efforts and fostering an appreciation for fungal diversity can contribute to more effective conservation outcomes.

Ex Situ Conservation:

Culture Collections: Ex situ conservation involves preserving fungal species outside their natural habitats, often in culture collections or gene banks. These collections maintain living cultures of fungi for research, education, and potential reintroduction into the wild. Maintaining genetic diversity in culture collections ensures that rare or endangered fungal species are preserved and available for future conservation efforts.

Seed Banks and Cryopreservation: Cryopreservation techniques can be used to store fungal spores and mycelium at low temperatures, ensuring their long-term viability. Seed banks and cryopreservation facilities play a role in conserving fungal genetic diversity and providing resources for research and restoration projects.

The conservation of fungal biodiversity is essential for maintaining ecosystem health and functionality. Effective conservation strategies involve understanding and documenting fungal diversity, addressing threats such as habitat destruction and climate change, implementing protected areas and restoration projects, raising public awareness, and utilizing ex situ conservation techniques. By prioritizing fungal conservation, we can protect the ecological roles and contributions of fungi to ecosystems and ensure the preservation of their rich and diverse biodiversity.

6.4.Conclusion

The intricate relationship between fungi and soil health is fundamental to the functioning and sustainability of natural ecosystems. As keystone organisms, fungi play crucial roles in maintaining soil structure, nutrient cycling, and ecosystem stability. Through their diverse forms and ecological functions, fungi contribute to the decomposition of organic matter, the formation of soil aggregates, and the recycling of essential nutrients such as carbon, nitrogen, and phosphorus. In forests, grasslands, and other ecosystems, fungi facilitate nutrient uptake for plants, support plant health through mycorrhizal associations, and enhance soil fertility and structure. Their ability to adapt to various environmental conditions, including extreme climates and nutrient-poor soils, underscores their resilience and importance in sustaining ecosystem health.

In conclusion, the conservation and understanding of fungal biodiversity are essential for preserving the integrity of natural ecosystems and ensuring their resilience to environmental changes. The interplay between fungi and soil health highlights the need for continued research, monitoring, and conservation efforts to protect these vital organisms and their habitats. By recognizing the indispensable roles fungi play in soil health and ecosystem functioning, we can better appreciate their contributions and implement strategies to safeguard their diversity and ecological functions. Addressing the challenges posed by habitat loss, climate change, and pollution will be critical for maintaining the balance and sustainability of natural ecosystems, ultimately ensuring the continued benefits that fungi provide to both the environment and human societies

Chapter 7

Climate Change and Soil Fungi

Climate change poses significant challenges to soil fungi, impacting their distribution, diversity, and ecological functions. Rising temperatures, altered precipitation patterns, and increased frequency of extreme weather events can disrupt the delicate balance within soil ecosystems. Elevated temperatures can influence fungal growth and metabolic rates, potentially leading to shifts in fungal community composition. For instance, some fungi may benefit from warmer conditions, expanding their ranges and potentially outcompeting other species. Conversely, other fungi, particularly those adapted to cooler environments, may experience reduced growth or even local extinctions if temperatures exceed their tolerance thresholds. Changes in precipitation can affect soil moisture levels, influencing fungal activity and distribution. For example, prolonged droughts may lead to reduced fungal biomass and altered nutrient cycling, while increased rainfall could promote the growth of moisture-loving fungi and affect soil structure.

Climate change can impact the interactions between soil fungi and other soil organisms, including plants, bacteria, and animals. Altered climate conditions may disrupt mycorrhizal associations, affecting plant nutrient uptake and overall plant health. These disruptions can have cascading effects on plant communities and ecosystem functions. Furthermore, climate-induced shifts in fungal communities may influence soil carbon storage and decomposition rates, with potential implications for global carbon cycling and greenhouse gas emissions. Understanding and predicting these impacts require integrated research efforts that combine field observations, experimental studies, and modeling approaches. By comprehensively assessing how climate change affects soil fungi and their roles in ecosystems, scientists can better anticipate potential changes and develop strategies to mitigate adverse effects and promote ecosystem resilience.

Climate change exerts multifaceted effects on soil fungi, significantly influencing their distribution, community composition, and ecological roles. Rising temperatures directly impact fungal physiology and growth rates, potentially altering fungal communities. As temperatures increase, certain fungi may thrive, leading to shifts in community dynamics and potentially displacing native species adapted to cooler conditions. Temperature-induced changes can also affect fungal reproductive cycles and sporulation rates, influencing their ability to colonize and persist in different habitats. Furthermore, warmer temperatures can enhance microbial activity in the soil, which might lead to accelerated decomposition of organic matter and alterations in nutrient cycling processes. This can result in changes to soil structure and fertility, which in turn affects plant growth and overall ecosystem health.

Precipitation patterns also play a crucial role in shaping fungal dynamics under climate change. Variations in rainfall can lead to changes in soil moisture, which influences fungal activity and distribution. Prolonged drought conditions may reduce fungal biomass and limit their ecological functions, such as decomposition and nutrient cycling. Conversely, increased precipitation can create conditions favorable for moisture-loving fungi, potentially leading to shifts in community composition and interactions with other soil organisms. Additionally, extreme weather events and increased frequency of heavy rainfall can lead to soil erosion and habitat disturbance, impacting fungal populations and their ecological roles. Understanding these climate-driven changes requires a comprehensive approach that includes monitoring fungal communities, assessing their responses to environmental changes, and modeling their potential impacts on ecosystem functions and services.

7.1. Impact of climate change on soil fungal populations and functions.

Climate change significantly affects soil fungal populations, altering their community structure and ecological functions. Rising temperatures can shift fungal distributions by favoring species that thrive in warmer conditions while disadvantaging those adapted to cooler climates. This temperature-driven shift can lead to a decrease in fungal diversity and potentially disrupt complex soil microbial networks. For instance, thermophilic fungi may become more prevalent, changing nutrient cycling processes and interactions with other soil organisms. Additionally, increased temperatures can accelerate fungal metabolic rates and growth, potentially leading to faster decomposition of organic matter. While this might enhance nutrient availability in some cases, it can also reduce soil organic carbon storage and affect long-term soil fertility.

Precipitation changes further compound these effects, influencing soil moisture and fungal activity. Variations in rainfall can lead to either drought conditions or waterlogging, both of which impact fungal populations differently. Drought conditions may reduce fungal biomass and inhibit their ability to decompose organic material, while excessive moisture can promote the growth of moisture-loving fungi and contribute to fungal diseases or pathogenic outbreaks. Such changes in soil moisture can also affect fungal interactions with plant roots, disrupting mycorrhizal associations crucial for plant nutrient uptake. These shifts in fungal populations and functions can cascade through ecosystems, affecting plant health, soil structure, and nutrient cycling, ultimately impacting ecosystem resilience and productivity. Understanding and mitigating these impacts requires integrated research efforts and adaptive management strategies to ensure the stability and sustainability of soil ecosystems under changing climate conditions.

Impact of Climate Change on Soil Fungal Populations

Climate change significantly affects soil fungal populations through its influence on temperature, precipitation patterns, and the frequency of extreme weather events. These environmental changes can alter fungal community composition, diversity, and distribution, with substantial implications for soil ecosystem

Temperature Changes:

Thermal Stress and Distribution Shifts: Rising temperatures can affect fungal growth rates, metabolic activity, and survival. Many soil fungi have specific temperature ranges for optimal growth, and deviations from these ranges can stress fungal populations. As temperatures increase, fungi adapted to cooler climates may struggle to survive, leading to shifts in community composition. In contrast, thermophilic fungi that thrive in warmer conditions may become more prevalent. This shift can disrupt existing fungal communities and reduce overall biodiversity, affecting the balance of ecological functions they perform, such as decomposition and nutrient cycling.

Extended Growing Seasons: Warmer temperatures can lengthen the growing season for many fungal species, potentially leading to increased fungal biomass and activity. However, this can also favor certain fast-growing or opportunistic fungi over more specialized or slower-growing species, further altering community dynamics. This extended growth period might lead to increased fungal competition, which can impact the diversity and stability of fungal communities.

Precipitation Patterns:

Moisture Variability: Changes in precipitation can lead to extreme soil moisture conditions, affecting fungal populations differently. Prolonged drought conditions can lead to reduced soil moisture, which can inhibit fungal growth and decrease fungal biomass. Drought stress can also reduce fungal diversity as some species are more drought-tolerant than others. Conversely, increased rainfall or waterlogging can create favorable conditions for moisture-loving fungi, potentially leading to shifts in community composition and increased fungal biomass. These changes can impact soil health and the functions fungi perform, such as organic matter decomposition and nutrient cycling.

Erosion and Soil Disturbance: Extreme precipitation events, such as heavy rainfall and flooding, can cause soil erosion and disrupt fungal habitats. Erosion can lead to the loss of topsoil and fungal populations, while flooding can alter soil structure and water availability, further impacting fungal communities. Such disturbances can lead to shifts in fungal populations and their functions, affecting overall soil ecosystem health.

Extreme Weather Events:

Impact on Fungal Survival: Increased frequency and intensity of extreme weather events, such as storms and heatwaves, can disrupt fungal populations by causing habitat destruction, soil erosion, and changes in moisture levels. Extreme events can directly damage fungal structures, reduce fungal biomass, and alter community composition. The frequency of such events can exacerbate the impact of other climate-related changes, leading to more pronounced shifts in fungal populations.

Cascading Effects: Extreme weather events can have cascading effects on soil ecosystems, impacting not only fungal populations but also their interactions with plants and other soil organisms. For example, extreme weather can influence the availability of organic matter and nutrients, affecting fungal growth and activity. This, in turn, can impact plant health and soil nutrient cycling, creating feedback loops that further influence fungal populations and community dynamics.

Climate change impacts soil fungal populations by altering temperature, precipitation, and extreme weather patterns. These changes can lead to shifts in fungal community composition, reduced diversity, and altered ecological functions. Understanding these impacts is crucial for managing soil health and ecosystem sustainability in the face of climate change.

Impact of Climate Change on Soil Fungal Functions

Climate change profoundly impacts the functions that soil fungi perform within ecosystems, affecting processes crucial for soil health and ecosystem sustainability. Soil fungi are integral to various ecological functions, including nutrient cycling, organic matter decomposition, and soil structure formation. Changes in climate, such as elevated temperatures, altered precipitation patterns, and increased frequency of extreme weather events, can significantly disrupt these functions in several ways:

Decomposition and Nutrient Cycling:

Temperature Effects: Rising temperatures can accelerate fungal metabolism and growth rates, which might lead to faster decomposition of organic matter. While this can temporarily increase nutrient availability, it may also lead to the rapid loss of soil organic carbon, affecting long-term soil fertility and carbon sequestration. Conversely, excessive heat can also stress fungi, potentially reducing their efficiency in breaking down organic materials. This imbalance can disrupt nutrient cycling and soil fertility, affecting plant growth and ecosystem productivity.

Moisture Effects: Changes in soil moisture due to altered precipitation can have contrasting impacts on fungal decomposition activity. Drought conditions can decrease fungal biomass and enzymatic activity, slowing down decomposition rates and leading to the accumulation of undecomposed organic matter. On the other hand, excessive moisture or waterlogging can enhance the growth of moisture-loving fungi, potentially increasing the decomposition of some organic substrates but also leading to reduced soil aeration and increased anaerobic conditions.

This can result in the accumulation of methane, a potent greenhouse gas, and alter the balance of soil microbial communities.

Soil Structure and Aggregation:

Temperature and Moisture Effects: Soil fungi contribute to soil structure and aggregation by producing hyphal networks that bind soil particles together, enhancing soil stability and porosity. Changes in temperature and moisture can affect fungal mycelial networks and their ability to form stable soil aggregates. For instance, prolonged droughts can reduce fungal activity and mycelial growth, leading to decreased soil aggregation and increased susceptibility to erosion. Excessive moisture can lead to soil compaction and loss of structural integrity due to the proliferation of certain fungi and the breakdown of organic matter into less stable forms.

Extreme Weather Events: Increased frequency of extreme weather events, such as heavy rainfall or storms, can disrupt fungal networks and soil structure. Heavy rainfall can cause soil erosion and waterlogging, which may damage fungal hyphae and reduce their capacity to form and maintain soil aggregates. This disruption can lead to decreased soil fertility, increased erosion, and altered soil hydrology, affecting overall soil health and ecosystem stability.

Plant-Fungal Interactions:

Mycorrhizal Associations: Soil fungi form symbiotic relationships with plant roots through mycorrhizal associations, aiding in nutrient uptake, particularly phosphorus, and enhancing plant health. Climate change can disrupt these interactions by altering soil conditions and affecting fungal colonization and activity. For example, temperature extremes and changes in soil moisture can impact the efficiency of mycorrhizal fungi in facilitating nutrient uptake. This can lead to reduced plant growth, increased susceptibility to diseases, and diminished ecosystem productivity.

Pathogen Dynamics: Climate change can also influence fungal pathogens that affect plant health. Warmer temperatures and increased humidity may favor the growth of pathogenic fungi, leading to increased incidences of plant diseases. This can have cascading effects on plant communities, crop yields, and ecosystem stability. Additionally, shifts in fungal pathogen populations can impact the effectiveness of existing plant-fungal interactions, further complicating plant-fungal dynamics.

Climate change has far-reaching effects on the functional roles of soil fungi within ecosystems. By altering temperature, moisture, and extreme weather patterns, climate change impacts decomposition rates, nutrient cycling, soil structure, and plant-fungal interactions. These changes can disrupt the delicate balance of soil ecosystems, affecting soil health, plant growth, and overall ecosystem functionality. Understanding these impacts is crucial for developing strategies to mitigate the adverse effects of climate change on soil fungi and ensuring the resilience and sustainability of ecosystems.

7.2. Fungi's role in carbon sequestration and mitigating global warming.

Fungi play a crucial role in carbon sequestration, significantly contributing to the global carbon cycle and helping mitigate global warming. Soil fungi are integral to the decomposition of organic matter, breaking down plant residues, leaf litter, and other organic materials into simpler compounds. This decomposition process not only recycles nutrients but also influences carbon storage in soils. Fungi contribute to the formation of soil organic matter by stabilizing carbon in the form of fungal biomass and fungal-generated soil aggregates. Their mycelial networks can physically protect carbon compounds from decomposition by binding soil particles together, thus enhancing the formation of stable soil organic carbon pools. These pools can store carbon for extended periods, potentially sequestering significant amounts of carbon dioxide (CO_2) from the atmosphere and reducing the greenhouse gas concentrations that drive global warming.

Additionally, fungi contribute to carbon sequestration through their role in symbiotic relationships with plants, particularly through mycorrhizal associations. Mycorrhizal fungi enhance plant nutrient uptake, which can lead to increased plant growth and greater input of organic carbon into the soil as plant roots exude carbon-rich compounds and leaf litter. This increased carbon input into the soil further promotes the development of fungal mycelium and soil organic matter, reinforcing the carbon sequestration process. By improving soil structure and stability, fungi also help maintain soil health, which supports ongoing carbon storage. Furthermore, fungi can influence the decomposition rate of organic matter; their presence can lead to slower decomposition rates and longer-term carbon storage in soils. Understanding and harnessing the role of fungi in carbon sequestration is crucial for developing effective strategies to mitigate global warming and enhance soil carbon storage, thereby contributing to climate change mitigation efforts.

Fungi's Role in Carbon Sequestration

Fungi play a pivotal role in carbon sequestration through several key processes that significantly impact the global carbon cycle. Their involvement in carbon sequestration is essential for maintaining soil health and mitigating climate change. Here's a detailed look at how fungi contribute to this process:

Decomposition and Formation of Soil Organic Matter:

Decomposition: Soil fungi are primary decomposers in terrestrial ecosystems. They break down complex organic materials such as plant residues, leaf litter, and wood into simpler forms. This decomposition process releases carbon in the form of carbon dioxide (CO_2) and other gases, but it also converts a portion of the carbon into more stable forms that can contribute to long-term soil carbon storage. Fungi utilize a range of enzymatic systems to decompose recalcitrant organic matter, including lignin and cellulose, which are resistant to decay by bacteria alone.

Formation of Soil Organic Matter: As fungi decompose organic matter, they contribute to the formation of soil organic matter (SOM). Fungal hyphae, which are the thread-like structures of fungi, bind soil particles together, forming aggregates that protect organic carbon from rapid decomposition. This process leads to the accumulation of stable soil organic carbon (SOC) that can persist in soils for decades or even centuries. The mycelium of fungi can also stabilize carbon in the soil by forming complex interactions with soil minerals, which helps to lock carbon into the soil matrix and reduces its susceptibility to decomposition.

Mycorrhizal Associations and Carbon Input:

Enhanced Plant Growth: Mycorrhizal fungi form symbiotic relationships with plant roots, significantly enhancing plant nutrient uptake, especially phosphorus. This relationship improves plant growth and productivity, leading to increased biomass and higher carbon inputs into the soil. Plants with mycorrhizal associations allocate more carbon to their roots and exude more carbon-rich compounds into the soil. This additional carbon input contributes to the development of fungal mycelium and increases soil organic matter.

Carbon Transfer to Soil: Mycorrhizal fungi facilitate the transfer of carbon from plants to the soil. The fungi intercept and utilize carbon compounds released by plant roots, and in return, provide essential nutrients to the plants. This exchange results in the incorporation of carbon into the fungal biomass and soil organic matter. Moreover, mycorrhizal fungi can increase the rate at which plant residues are incorporated into the soil, promoting the formation of stable organic carbon pools.

Stabilization and Protection of Soil Carbon:

Soil Aggregation: Fungal hyphae play a crucial role in soil aggregation by physically binding soil particles together. This aggregation process leads to the formation of soil aggregates that protect organic carbon from microbial decomposition and erosion. Soil aggregates create microenvironments where carbon can be stored more securely, thus enhancing the long-term sequestration of carbon in the soil.

Physical and Chemical Interactions: Fungi also interact with soil minerals, such as clay and silt particles, through their mycelium. These interactions can form stable complexes that protect organic carbon from decomposition. By binding carbon compounds with soil minerals, fungi contribute to the formation of stable soil organic carbon pools that persist in the soil for extended periods.

Fungi contribute to carbon sequestration through their roles in decomposition, soil organic matter formation, and mycorrhizal associations. Their ability to stabilize carbon in soil aggregates and interact with soil minerals helps lock carbon into the soil, mitigating greenhouse gas emissions and supporting soil health. Understanding these processes is crucial for developing strategies to enhance carbon sequestration and combat climate change.

Fungi's Role in Mitigating Global Warming

Fungi play a multifaceted role in mitigating global warming through their involvement in carbon sequestration, nutrient cycling, and ecosystem stability. Their activities significantly influence the carbon cycle and soil health, contributing to efforts aimed at reducing greenhouse gas concentrations and enhancing environmental resilience. Here's a detailed explanation of how fungi contribute to mitigating global warming:

Carbon Sequestration:

Decomposition and Soil Organic Matter Formation: Fungi are crucial decomposers of organic matter, such as plant residues, leaf litter, and wood. By breaking down these materials, fungi convert carbon into forms that can be stored in the soil as soil organic matter (SOM). Fungi decompose complex organic compounds, including lignin and cellulose, which are resistant to bacterial decomposition. This process not only releases carbon dioxide (CO_2) but also helps in forming stable carbon compounds that persist in the soil for long periods. Fungal hyphae bind soil particles together to form aggregates, which protect organic carbon from rapid decomposition and erosion. This stabilization leads to the long-term storage of carbon in soils, reducing atmospheric CO_2 levels.

Mycorrhizal Associations: Mycorrhizal fungi form symbiotic relationships with plant roots, enhancing plant nutrient uptake and growth. This relationship leads to increased plant biomass and greater carbon inputs into the soil. Plants with mycorrhizal associations allocate more carbon to their roots and release more carbon-rich exudates into the soil. This additional carbon is then utilized by fungi and incorporated into soil organic matter, further contributing to carbon sequestration. Mycorrhizal fungi also facilitate the transfer of carbon from plants to the soil, enhancing the formation of stable soil carbon pools.

Mitigation of Greenhouse Gas Emissions:

Influence on Methane Dynamics: Soil fungi influence methane (CH_4) dynamics, a potent greenhouse gas. Certain fungi can act as methanotrophs, microorganisms that consume methane as a carbon source. By metabolizing methane, these fungi help reduce atmospheric methane concentrations, contributing to greenhouse gas mitigation. Additionally, fungal activity in soils can influence the balance between methane production and consumption, affecting the overall emissions of this gas from soils.

Nitrate Reduction and Nitrous Oxide Emissions: Fungi play a role in nitrogen cycling, including the reduction of nitrates in the soil. This process can affect the production of nitrous oxide (N_2O), another potent greenhouse gas. By influencing nitrogen transformations, fungi can impact N_2O emissions, helping to mitigate greenhouse gas concentrations in the atmosphere.

Enhancing Soil Health and Resilience:

Soil Structure and Stability: Fungi contribute to soil structure by forming hyphal networks that bind soil particles into aggregates. This enhances soil porosity, water infiltration, and resistance to erosion. Improved soil structure supports plant growth and resilience, contributing to ecosystem stability and reducing the impacts of extreme weather events. Healthy soils with robust fungal networks can better retain carbon and nutrients, supporting long-term carbon sequestration and enhancing ecosystem resilience to climate change.

Ecosystem Services: By supporting plant health, improving nutrient availability, and enhancing soil structure, fungi provide critical ecosystem services that contribute to climate change mitigation. Healthy, well-functioning ecosystems with diverse fungal communities can better withstand environmental stressors, including those induced by climate change. This resilience helps maintain ecosystem services, such as carbon sequestration, water regulation, and nutrient cycling, which are essential for mitigating global warming.

Biodiversity and Ecosystem Function:

Support for Biodiversity: Fungi contribute to biodiversity by interacting with a wide range of plant and microbial species. Diverse fungal communities support complex ecosystem functions, including carbon cycling, nutrient availability, and soil structure formation. Maintaining fungal biodiversity enhances ecosystem stability and resilience, which is crucial for climate change mitigation.

Ecosystem Functioning: The diverse functions performed by fungi, including decomposition, nutrient cycling, and soil aggregation, are essential for maintaining healthy ecosystems. These functions contribute to carbon sequestration and mitigate greenhouse gas emissions, supporting efforts to address global warming.

Fungi play a vital role in mitigating global warming through carbon sequestration, greenhouse gas regulation, and the enhancement of soil health and ecosystem resilience. Their activities in decomposition, mycorrhizal associations, and nutrient cycling contribute to the long-term storage of carbon in soils, reduction of greenhouse gas emissions, and support for ecosystem functions. Understanding and leveraging the role of fungi in climate change mitigation is crucial for developing effective strategies to combat global warming and enhance environmental sustainability.

7.3. The potential effects of shifting fungal populations on ecosystems.

The shifting of fungal populations due to environmental changes, including climate change, land use alterations, and pollution, can have profound effects on ecosystems. Fungi play critical roles in nutrient cycling, soil structure formation, and plant health, and changes in their populations can disrupt these essential functions. As fungal communities shift, there may be a reduction in fungal diversity, which can lead to diminished ecosystem resilience and functionality. For example, the loss of specific fungal species that contribute to nutrient cycling could impair the decomposition of organic matter, resulting in decreased soil fertility and altered nutrient availability for plants. Similarly, changes in mycorrhizal fungal populations, which form symbiotic relationships with plant roots, can impact plant growth and health, potentially leading to reduced crop yields and altered plant community composition. The disruption of fungal communities can also affect soil structure, as certain fungi contribute to soil aggregation and stabilization. A decline in these fungi may lead to increased soil erosion, reduced water infiltration, and decreased soil carbon storage. Furthermore, the introduction of non-native or invasive fungal species can outcompete native fungi, leading to shifts in ecosystem dynamics and potential loss of ecosystem services. Overall, the potential effects of shifting fungal populations on ecosystems are multifaceted and can have cascading impacts on soil health, plant productivity, and overall ecosystem stability. The shifting of fungal populations in ecosystems due to various factors, including climate change, land use changes, and pollution, can have significant and multifaceted effects on ecosystem structure and function. Here's a detailed explanation of these potential effects:

Disruption of Nutrient Cycling:

Decomposition Processes: Fungi are primary decomposers of organic matter, including plant litter, wood, and other organic materials. Shifts in fungal populations can alter the rates and efficiency of decomposition processes. For example, if fungal species specialized in breaking down complex compounds like lignin and cellulose are reduced, the decomposition of these materials may slow down, leading to an accumulation of undecomposed organic matter. This can affect nutrient release and cycling, potentially reducing soil fertility and altering the availability of essential nutrients for plants.

Nutrient Availability: Mycorrhizal fungi, which form symbiotic relationships with plant roots, play a crucial role in nutrient uptake, particularly phosphorus. Changes in mycorrhizal populations can affect nutrient availability for plants, leading to changes in plant growth and productivity. If beneficial mycorrhizal fungi are reduced, plants may experience nutrient deficiencies, which can impact their health, growth, and competitive ability in the ecosystem.

Altered Soil Structure and Health:

Soil Aggregation: Fungal hyphae contribute to soil aggregation by binding soil particles together, which improves soil structure and stability. Shifts in fungal populations can affect soil aggregation, leading to changes in soil porosity, water infiltration, and erosion susceptibility. For instance, a decline in fungi that form soil aggregates can result in increased soil erosion, reduced water-holding capacity, and compromised soil health.

Soil Carbon Storage: Fungi play a role in stabilizing soil organic carbon by incorporating it into fungal biomass and soil aggregates. Changes in fungal populations can influence the rate of carbon sequestration in soils. If fungi that contribute to carbon stabilization are diminished, soil carbon storage may decrease, leading to higher atmospheric carbon dioxide levels and contributing to global warming.

Impacts on Plant Communities:

Plant Health and Productivity: Mycorrhizal fungi enhance plant nutrient uptake and can affect plant health and productivity. Shifts in mycorrhizal populations can lead to changes in plant community composition and productivity. For example, if certain mycorrhizal fungi that benefit specific plant species are reduced, those plants may struggle with nutrient deficiencies, potentially leading to shifts in plant community dynamics and reduced biodiversity.

Invasive Species: Changes in fungal communities can create opportunities for invasive fungal species to establish and spread. Invasive fungi can outcompete native species, disrupt established plant-fungal interactions, and alter ecosystem functions. This can lead to reduced biodiversity, changes in plant community composition, and potentially harmful effects on ecosystem stability.

Effects on Ecosystem Services:

Water Regulation: Fungi contribute to soil structure and water regulation by enhancing soil aggregation and water-holding capacity. Shifts in fungal populations can affect these functions, leading to altered water infiltration, increased runoff, and higher susceptibility to flooding or drought conditions.

Disease Regulation: Soil fungi can influence plant disease dynamics by competing with pathogenic microorganisms and suppressing their growth. Changes in fungal populations can affect this balance, potentially leading to increased disease pressure on plants and reduced crop yields.

Interactions with Other Organisms:

Microbial Interactions: Fungi interact with a wide range of soil microorganisms, including bacteria and other fungi. Shifts in fungal populations can alter these interactions, impacting microbial community structure and function. This can have cascading effects on soil health, nutrient cycling, and overall ecosystem dynamics.

Animal Relationships: Fungi also play a role in interactions with soil-dwelling animals, such as insects and mammals, which may rely on fungi for food or habitat. Changes in fungal populations can impact these relationships, affecting the broader ecosystem web.

The potential effects of shifting fungal populations on ecosystems are diverse and profound. They can disrupt nutrient cycling, alter soil structure and health, impact plant communities and ecosystem services, and influence interactions with other organisms. Understanding these effects is crucial for managing ecosystems and mitigating the impacts of environmental changes on fungal populations and overall ecosystem health.

7.4.Conclusion

Climate change presents significant challenges to soil fungal communities, impacting their diversity, distribution, and ecological functions. As global temperatures rise and precipitation patterns shift, fungal populations are responding in complex ways, which can have cascading effects on soil health and ecosystem stability. Changes in temperature and moisture levels can alter fungal growth rates, reproduction, and interactions with other soil organisms. This can lead to shifts in fungal community composition, potentially reducing the diversity of beneficial fungi and favoring pathogenic or invasive species. Such shifts can disrupt key ecosystem processes, including nutrient cycling, soil structure formation, and carbon sequestration. The resulting alterations in soil health and plant productivity underscore the need for a deeper understanding of how climate change affects soil fungi and the ecosystems they support.

Addressing the impacts of climate change on soil fungi requires a multifaceted approach that includes monitoring fungal populations, understanding their responses to environmental changes, and implementing strategies to mitigate potential negative effects. Conservation efforts should focus on preserving fungal diversity and enhancing soil resilience through sustainable land management practices. Additionally, integrating fungal ecology into climate change adaptation strategies can help ensure the continued provision of essential ecosystem services, such as nutrient cycling and carbon sequestration. By fostering resilience in fungal communities and maintaining healthy soil ecosystems, we can better manage the impacts of climate change and support broader environmental sustainability goal

Chapter 8

Methods of Studying Soil Fungi

Studying soil fungi involves a range of techniques that help researchers understand their diversity, distribution, and ecological roles. Traditional methods for studying soil fungi include morphological and culturing techniques, which provide foundational insights into fungal identification and characterization. Morphological techniques involve examining fungal structures under a microscope, including spores, hyphae, and fruiting bodies, to identify fungal species based on their physical traits. This approach often requires the isolation of fungi from soil samples, which is done using selective media that supports the growth of specific fungal groups while inhibiting others. By culturing fungi in the laboratory, researchers can observe their growth patterns, reproductive structures, and biochemical properties, contributing to taxonomic classification and ecological understanding. However, these methods are limited by their reliance on cultivable fungi, potentially overlooking a significant portion of the soil fungal diversity.

In recent years, molecular techniques have revolutionized the study of soil fungi by allowing for more comprehensive and detailed analysis. DNA-based methods, such as polymerase chain reaction (PCR), sequencing, and metagenomics, enable researchers to identify fungal species directly from soil samples without the need for culturing. High-throughput sequencing technologies, such as next-generation sequencing (NGS), allow for the analysis of fungal DNA sequences from soil, providing insights into the diversity and community composition of fungi. This approach can detect both known and novel fungal species, offering a more complete picture of soil fungal diversity. Molecular methods also facilitate the study of functional genes and metabolic pathways, enhancing our understanding of the ecological roles of different fungal species in nutrient cycling, organic matter decomposition, and interactions with plants and other soil organisms.

Ecological and field-based approaches complement laboratory and molecular techniques by examining the interactions between soil fungi and their environment. Soil surveys and sampling involve collecting soil from various locations and depths to assess fungal diversity and distribution patterns across different habitats and conditions. Environmental parameters such as soil moisture, temperature, pH, and organic matter content are often measured to understand their influence on fungal communities. Experimental manipulations, such as altering soil conditions or introducing specific fungal species, help researchers study the effects of environmental changes on fungal activity and ecosystem functions. Additionally, the use of stable isotope tracers and radio-labeled compounds can track fungal contributions to nutrient cycling and carbon sequestration in situ. Combining these approaches provides a holistic view of soil fungal dynamics and their role in ecosystem processes, offering valuable insights for conservation and land management practices.

8.1.Techniques for identifying and characterizing soil fungi (microscopy, molecular methods).

Identifying and characterizing soil fungi involves a combination of techniques that leverage morphological, molecular, and biochemical methods to provide a comprehensive understanding of fungal diversity, taxonomy, and ecological functions. Techniques for identifying and characterizing soil fungi are multifaceted, incorporating both traditional and modern methods to achieve comprehensive results. Initial identification often begins with soil sampling and culturing, where soil is plated onto selective media such as Potato Dextrose Agar (PDA) or Sabouraud Dextrose Agar (SDA) to isolate fungal colonies. These cultures are then examined for morphological traits like colony color, texture, and growth patterns, which provide preliminary identification clues. Microscopic examination further refines identification by revealing structural features such as spore morphology and hyphal characteristics. For more precise and comprehensive identification, molecular techniques such as Polymerase Chain Reaction (PCR) and DNA sequencing are employed to analyze genetic material, enabling the identification of fungi to the species level and beyond. Tools like Internal Transcribed Spacer (ITS) sequencing and whole-genome sequencing offer detailed insights into fungal taxonomy and function. Advanced bioinformatics software integrates and analyzes this genetic data, linking it to fungal databases for accurate identification and functional characterization. Together, these techniques provide a robust framework for studying soil fungi, enhancing our understanding of their diversity, ecological roles, and interactions within soil ecosystems. Here's a detailed explanation of these techniques:

1. Morphological Techniques

Morphological techniques for identifying and characterizing soil fungi involve detailed observation and analysis of fungal structures, both macroscopic and microscopic, to determine species and functional traits. Initially, soil samples are cultured on selective agar media, such as Potato Dextrose Agar (PDA) or Sabouraud Dextrose Agar (SDA), where fungi grow into visible colonies. These colonies are examined for distinctive macroscopic features, including color, texture, and growth patterns, which can provide initial clues about the fungal genus or species. Microscopic examination is then employed to assess more intricate details, such as hyphal structure, spore morphology, and reproductive structures (e.g., conidia, asci, basidia). Techniques like wet mounts, staining, and the use of compound microscopes allow for the visualization of these features, which are crucial for accurate identification. Additionally, specific morphological characteristics, such as the arrangement of spores or the presence of particular structures like fruiting bodies, can be compared to taxonomic keys and fungal identification guides. This traditional approach, though

124

labor-intensive, remains fundamental in fungal taxonomy and complements molecular techniques by providing a detailed morphological context that aids in the accurate classification and understanding of fungal diversity

Microscopy: Traditional fungal identification often begins with examining fungal structures under a microscope. Microscopy is a critical technique for identifying and characterizing soil fungi, offering detailed insights into their morphological and structural features. Using light microscopy, researchers observe fungi at various stages of development, including spore formation, hyphal growth, and reproductive structures. Techniques such as bright-field microscopy provide a general view of fungal structures, while differential interference contrast (DIC) microscopy enhances contrast and reveals finer details of fungal morphology. For more detailed analysis, fluorescence microscopy can be employed to highlight specific cellular components or structures using fluorescent dyes or markers. Advanced techniques like scanning electron microscopy (SEM) and transmission electron microscopy (TEM) offer high-resolution imaging, allowing for the examination of ultrastructural details, such as cell wall composition and internal organelles. These microscopy methods enable precise identification of fungal species based on unique morphological traits, support the study of fungal development and interactions, and contribute to our understanding of their ecological roles and functions in soil ecosystems. Key features include:

Spore Morphology: The shape, size, and color of fungal spores can help identify species. Techniques like light microscopy and electron microscopy provide detailed images of spore structures. Spore morphology is a key aspect of fungal identification and classification, focusing on the size, shape, color, and surface characteristics of fungal spores. Spores, which are the reproductive units of fungi, exhibit a wide range of morphological traits that are critical for distinguishing between different fungal species. Detailed observation under a microscope allows researchers to examine spore features such as the presence of ornamentation, the number of cells, and the arrangement of spore walls. For instance, the presence of distinct ridges, spines, or smooth surfaces can help differentiate between similar-looking fungi. Additionally, spore color and size can provide further taxonomic clues. Accurate characterization of spore morphology is essential for identifying fungal species, understanding their life cycles, and studying their roles in ecosystems. This approach, while often used in conjunction with molecular techniques, remains a fundamental method for fungal taxonomy and ecological research.

Hyphal Structure: Observing the arrangement and type of hyphae (e.g., septate or non-septate) and their growth patterns can aid in identification. Hyphal structure is a critical morphological feature for identifying and classifying fungi, as it provides insights into the organization and growth patterns of fungal mycelium. Hyphae, the filamentous structures that make up the main vegetative part of a fungus, can vary widely in appearance, including their branching patterns, septation (presence of cross-walls), and texture. Observing hyphal structure under a microscope allows researchers to assess features such as the presence of clamp connections, which are characteristic of certain fungal groups, and the degree of branching, which can aid in distinguishing between species. Additionally, the width and wall thickness of hyphae, as well as their pigmentation and the presence of specialized structures like rhizomorphs or sclerotia, can provide further taxonomic information. Detailed examination of hyphal structure is essential for accurate fungal identification, understanding fungal growth habits, and exploring their ecological roles and interactions with other organisms.

Fruit Bodies: For fungi that produce visible fruiting bodies (e.g., mushrooms), examining their macroscopic characteristics such as shape, size, color, and texture is crucial. Fruit bodies, or sporocarps, are the reproductive structures of fungi that are often visible above ground and are crucial for spore production and dispersal. They vary widely in form, size, and appearance across different fungal taxa, making them important for fungal identification and classification. The morphology of fruit bodies, including characteristics such as shape, color, texture, and size, can provide significant taxonomic clues. For instance, mushrooms, puffballs, and cup fungi each have distinct fruit body structures that aid in their identification. Features like gills, pores, or teeth on the underside of the cap, as well as the arrangement and type of spore-bearing surfaces, are key for differentiating species. Additionally, the development stage of fruit bodies can offer insights into the lifecycle and ecology of the fungus. Detailed observation of fruit bodies, often in conjunction with molecular and ecological data, enhances our understanding of fungal biodiversity, their roles in ecosystems, and their potential applications in fields such as medicine and agriculture.

Culturing: Isolating fungi from soil samples onto selective agar media helps in obtaining pure cultures for further study. Culturing is a fundamental technique for isolating and growing soil fungi under controlled laboratory conditions, enabling detailed study of their morphology, physiology, and biochemical properties. The process begins with the collection of soil samples, which are then diluted and plated onto selective or general-purpose agar media, such as Potato Dextrose Agar (PDA) or Sabouraud Dextrose Agar (SDA). These media provide nutrients that support fungal growth while inhibiting the proliferation of bacteria. Once plated, the cultures are incubated at optimal temperatures to encourage fungal colonization. Over time, individual fungal colonies develop, each displaying distinct morphological characteristics such as colony color, texture, and edge shape. These visible traits are used to make preliminary identifications. For further characterization, isolated colonies can be sub-cultured onto fresh media to obtain pure cultures, which are then examined microscopically to assess detailed structures like hyphal morphology, spore types, and reproductive features. Culturing is essential for studying the physiological and biochemical aspects of fungi, testing their responses to different environmental conditions, and conducting further molecular analyses to accurately identify species and explore their ecological roles. This involves:

Media Selection: Using media that support the growth of specific fungal groups (e.g., potato dextrose agar for general fungi, Sabouraud dextrose agar for yeasts). Media selection is a critical aspect of fungal culture and isolation, involving the choice of nutrient-rich substrates that support the growth of specific fungi while suppressing unwanted microorganisms. The type of media used depends on the fungal species being studied and the goals of the research. For example, general-purpose media like Potato Dextrose Agar (PDA) or Sabouraud Dextrose Agar (SDA) are commonly used for culturing a wide range of fungi, while selective media, such as Rose Bengal Chloramphenicol Agar, can inhibit the growth of bacteria and facilitate the isolation of fungal species. Media can also be supplemented with specific nutrients or inhibitors to promote the growth of particular fungi or to encourage the expression of certain traits, such as enzyme production. Proper media selection ensures optimal growth conditions for fungi, accurate isolation of species, and reliable data on fungal physiology, diversity, and ecological roles. This foundational step is essential for successful fungal research and applications in agriculture, environmental management, and biotechnology.

Identification: Observing growth patterns, colony morphology, and reproductive structures (like conidia or sporangia) under the microscope. Identification of soil fungi involves determining the specific species or genera present in a given sample, utilizing a combination of morphological, molecular, and biochemical techniques. Morphological identification typically relies on observing fungal growth characteristics, such as colony morphology, spore shape, and reproductive structures, under a microscope. However, for more precise and comprehensive identification, molecular techniques such as DNA sequencing and PCR-based methods are employed. Molecular approaches use genetic markers, such as ribosomal RNA genes or specific genomic regions, to distinguish between species and provide insights into their genetic diversity. Additionally, biochemical tests can assess the production of specific metabolites or enzyme activities unique to certain fungi. Combining these methods enhances the accuracy of fungal identification, supports the understanding of fungal biodiversity, and facilitates research on their ecological roles, interactions, and potential applications in various fields.

2. Molecular Techniques

Molecular techniques have significantly advanced the study of soil fungi by providing detailed genetic and functional insights that complement traditional morphological methods. Polymerase Chain Reaction (PCR) is a fundamental molecular technique used to amplify specific fungal DNA regions, such as the Internal Transcribed Spacer (ITS) regions, which are critical for accurate species identification and taxonomic classification. DNA sequencing technologies, including Sanger sequencing and next-generation sequencing (NGS), enable the sequencing of entire fungal genomes or specific genes, revealing comprehensive genetic information and facilitating the identification of fungi at a species or strain level. Molecular techniques such as quantitative PCR (qPCR) and reverse-transcription PCR (RT-PCR) allow for the quantification of fungal

DNA and the assessment of gene expression, respectively, providing insights into fungal activity and interactions. Additionally, metagenomics, which involves sequencing all the genetic material in a soil sample, allows researchers to analyze fungal communities without culturing, revealing diversity and functional potentials in natural ecosystems. Bioinformatics tools are used to analyze sequencing data, compare genomes, and interpret functional annotations, thereby enhancing our understanding of fungal biology and their roles in soil ecosystems. These molecular methods offer precision and depth in fungal research, uncovering the genetic basis of ecological functions, interactions, and adaptations.

Polymerase Chain Reaction (PCR): PCR amplifies specific fungal DNA regions, facilitating the identification of fungi at the species or genus level. Polymerase Chain Reaction (PCR) is a powerful and widely used molecular technique that amplifies specific DNA sequences, making it essential for studying soil fungi. By targeting particular regions of fungal DNA, such as the Internal Transcribed Spacer (ITS) regions, PCR enables researchers to detect, identify, and quantify fungal species with high precision. The process involves three main steps: denaturation, where the double-stranded DNA is heated to separate it into single strands; annealing, where primers specific to the target DNA sequence bind to the complementary regions on the single-stranded DNA; and extension, where a DNA polymerase enzyme synthesizes new DNA strands by extending from the primers. This cyclical process results in the exponential amplification of the target DNA sequence, producing millions of copies from a small initial sample. PCR is particularly valuable for identifying fungi in complex soil samples, studying fungal biodiversity, and examining gene expression. It is complemented by techniques such as gel electrophoresis, which separates and visualizes the amplified DNA fragments, and quantitative PCR (qPCR), which measures the amount of DNA or RNA to assess fungal abundance and activity. Techniques include:

Species-Specific Primers: Primers designed for specific fungal taxa can identify known species. Species-specific primers are tailored DNA sequences used in polymerase chain reaction (PCR) to selectively amplify genetic material from a particular fungal species, allowing for precise identification and quantification. These primers are designed to bind to unique regions of the target species' genome, ensuring that only the DNA of the specific fungus is amplified, while other fungal or soil DNA remains unaffected. In soil fungal studies, species-specific primers are invaluable for detecting and studying rare or economically significant fungi, monitoring their abundance in environmental samples, and assessing their roles within microbial communities. By targeting unique genetic markers, researchers can gain accurate insights into fungal diversity, distribution, and ecological functions, facilitating studies on fungal interactions with plants, soil health, and ecosystem dynamics. This targeted approach enhances the reliability of molecular analyses and contributes to a deeper understanding of fungal species' ecological roles and their impacts on soil environments.

Universal Primers: Broad-range primers (e.g., ITS1/ITS4) amplify ribosomal DNA regions common to many fungi, useful for general fungal community analysis. Universal primers are designed to amplify a broad range of fungal species by targeting conserved regions of the fungal genome that are shared across multiple taxa. These primers are particularly useful for capturing genetic material from diverse fungal communities in environmental samples, allowing researchers to identify and study a wide variety of fungi without needing species-specific primers for each one. Universal primers typically bind to highly conserved regions of the ribosomal RNA genes, such as the 18S or ITS regions, which are present in many fungal species. By using universal primers in techniques like PCR and subsequent sequencing, researchers can profile fungal biodiversity, assess community composition, and explore functional potentials in soil environments. This approach is especially valuable for metagenomic studies and surveys of fungal diversity, providing a comprehensive overview of fungal populations and their ecological roles within ecosystems.

Sequencing: DNA sequencing provides detailed information about fungal species and their genetic makeup. Sequencing is a crucial molecular technique used to determine the exact order of nucleotides in a DNA or RNA molecule, providing comprehensive insights into fungal genomes and metagenomes. Next-Generation Sequencing (NGS) technologies, such as Illumina sequencing and Oxford Nanopore sequencing, have revolutionized fungal research by enabling high-throughput, cost-effective sequencing of entire genomes or specific genetic regions. These technologies generate vast amounts of sequencing data, which can be used to identify fungal species, explore genetic diversity, and elucidate functional potentials. In fungal studies, sequencing of regions like the Internal Transcribed Spacer (ITS) allows for precise taxonomic identification, while whole-genome sequencing reveals the complete genetic makeup of fungal species, providing insights into their metabolic pathways, adaptation mechanisms, and ecological roles. Additionally, RNA sequencing (RNA-Seq) analyzes transcriptomes to assess gene expression patterns and functional responses to environmental conditions. The extensive data generated through sequencing is analyzed using bioinformatics tools to assemble genomes, annotate genes, and interpret functional elements, thereby enhancing our understanding of fungal biology, ecology, and their interactions within soil ecosystems. Methods include:

Sanger Sequencing: Used for sequencing specific genes, such as the internal transcribed spacer (ITS) region of rDNA, to identify fungal species. Sanger sequencing, also known as chain-termination sequencing, is a widely used method for determining the nucleotide sequence of DNA. This technique involves amplifying the DNA of interest, followed by sequencing it using fluorescently labeled dideoxy nucleotides that terminate DNA synthesis at specific points. As the DNA strands are extended and terminated, the resulting fragments are separated by size through capillary electrophoresis, and the sequence is read based on the emitted fluorescence. In soil fungal studies, Sanger sequencing is employed to accurately sequence specific genes, such as ribosomal RNA genes, which are essential for identifying fungal species and understanding their genetic diversity. This method is valued for its high accuracy and reliability, making it a fundamental tool for obtaining precise genetic information, studying fungal taxonomy, and investigating the genetic basis of fungal functions and interactions within soil ecosystems.

Next-Generation Sequencing (NGS): High-throughput sequencing technologies (e.g., Illumina, PacBio) enable the analysis of entire fungal communities by sequencing multiple DNA fragments simultaneously, revealing a comprehensive picture of fungal diversity. Next-Generation Sequencing (NGS) represents a revolutionary advancement in genomic technology, enabling high-throughput sequencing of DNA and RNA with exceptional speed and depth. Unlike traditional methods, NGS can sequence millions of DNA fragments simultaneously, providing comprehensive data on the entire genome or specific regions of interest. In the context of soil fungi, NGS is used to generate detailed profiles of fungal communities, identify novel fungal species, and explore functional gene expression. Techniques such as Illumina sequencing and nanopore sequencing allow researchers to capture vast amounts of genetic information, uncovering variations, mutations, and complex interactions within fungal genomes. NGS facilitates metagenomic studies, where entire soil samples are analyzed to understand the diversity and functional potential of fungal communities. This high-resolution approach enhances our ability to study fungal ecology, discover new fungal taxa, and investigate the roles of fungi in soil health and ecosystem functioning.

Metagenomics: This approach involves sequencing all the genetic material in a soil sample to study the fungal community structure and functional potential without needing to culture fungi. Metagenomics is a cutting-edge technique that analyzes the collective genomic material from environmental samples, offering a comprehensive view of microbial communities, including soil fungi, without the need for culturing. By sequencing all the DNA present in a soil sample, metagenomics captures the diversity and abundance of fungal species and their functional genes, providing insights into the complex interactions and ecological roles of fungi in their natural habitats. This approach involves extracting DNA from the soil, followed by high-throughput sequencing to generate large-scale data sets. Bioinformatics tools then analyze this data to identify fungal taxa, assess community structure, and predict functional capabilities. Metagenomics enables the study of fungal diversity at a much finer scale than traditional methods, revealing novel and rare species and uncovering their contributions to processes like nutrient cycling and soil health. It also helps to understand how fungal communities respond to environmental changes and disturbances, offering valuable insights into ecosystem dynamics and the impact of human activities on soil health. It helps in:

Diversity Profiling: Identifying the presence and abundance of various fungal taxa in complex soil environments. Diversity profiling is a comprehensive approach used to assess and characterize the variety of fungal species present in a given soil environment. This involves employing a range of techniques to catalog and quantify the different fungal taxa, including traditional methods like morphological identification and modern molecular techniques such as DNA sequencing and metagenomics. By analyzing fungal communities through methods like ribosomal RNA gene sequencing or fingerprinting techniques, researchers can obtain detailed profiles of fungal diversity, including species richness, abundance, and community structure. Diversity profiling helps in understanding the ecological roles of different fungi, their interactions with other soil microorganisms, and their contributions to ecosystem functions such as nutrient cycling and organic matter decomposition. This information is vital for studying the impact of environmental changes, land management practices, and conservation efforts on soil fungal communities, ultimately contributing to better soil health management and ecosystem sustainability.

Functional Gene Analysis: Exploring genes related to fungal functions, such as decomposition or nutrient uptake. Functional gene analysis is a crucial method for understanding the specific roles and activities of genes within soil fungi, providing insights into their biochemical and ecological functions. This technique involves studying genes that encode for key functional proteins, such as enzymes involved in nutrient cycling or secondary metabolite production. By using approaches such as quantitative PCR (qPCR), gene expression profiling, and transcriptomics, researchers can measure the activity levels of these genes under various environmental conditions or stressors. Functional gene analysis allows scientists to link gene expression with fungal processes, such as decomposition, nutrient uptake, or pathogen resistance, and to identify genes that are critical for these functions. This detailed understanding of gene functions and their regulation helps in elucidating how soil fungi contribute to soil health, ecosystem stability, and their interactions with plants and other microorganisms, offering potential applications in biotechnology and environmental management.

3. Biochemical Techniques

Biochemical techniques are instrumental in studying soil fungi by examining their metabolic activities, enzyme production, and interactions with the soil environment. Techniques such as enzyme assays and metabolite profiling provide insights into the biochemical processes fungi use to decompose organic matter, cycle nutrients, and interact with other microorganisms. Enzyme assays measure the activity of specific enzymes, such as cellulases or laccases, which are crucial for breaking down complex organic materials in the soil. Metabolite profiling involves analyzing the range of metabolites produced by fungi, which can reveal their metabolic pathways and ecological functions. Additionally, biochemical techniques like protein electrophoresis and mass spectrometry help identify and characterize fungal proteins, offering information on their roles in fungal physiology and interactions. These methods complement molecular and morphological techniques by providing a deeper understanding of the functional aspects of fungal biology, including how fungi contribute to soil health, nutrient cycling, and ecosystem stability.

Fatty Acid Methyl Ester (FAME) Analysis: This method profiles the fatty acid composition of fungal cell membranes. It can distinguish between fungal groups based on their unique fatty acid profiles. Fatty Acid Methyl Ester (FAME) analysis is a specialized biochemical technique used to profile and identify the fatty acids present in fungal cell membranes and other biological materials. In this method, fatty acids are first extracted from the fungal biomass and then converted into their methyl ester derivatives through a process called methylation. These FAMEs are then separated and quantified using gas chromatography (GC), which provides a detailed fatty acid profile based on the retention times and peak areas of the different esters. This analysis is valuable for studying fungal taxonomy and physiology, as the fatty acid composition can be distinctive and specific to different fungal species or strains. By comparing FAME profiles, researchers can identify and classify fungi, investigate their metabolic adaptations, and understand their ecological roles and interactions. FAME analysis also aids in assessing the nutritional and biochemical properties of fungi, contributing to our knowledge of their functional diversity and applications in fields such as agriculture and biotechnology.

Proteomics: Analyzing fungal proteins provides insights into fungal physiology and interactions. Proteomics is an advanced biochemical technique that focuses on the comprehensive analysis of the entire protein complement expressed by fungi, providing detailed insights into their functional biology, physiology, and interactions. In proteomics, proteins are extracted from fungal samples and subjected to techniques such as two-dimensional gel electrophoresis (2-DE) or liquid chromatography coupled with mass spectrometry (LC-MS) to separate, identify, and quantify them. These methods allow researchers to determine the abundance, modifications, and interactions of proteins within fungal cells, revealing how fungi respond to environmental changes, stressors, or nutrient availability. Proteomics can uncover novel proteins, elucidate metabolic pathways, and provide insights into fungal roles in soil ecosystems, such as their involvement in nutrient cycling and interactions with plants or other microorganisms. This technique enhances our understanding of fungal biology by linking gene expression with protein function and helps in identifying biomarkers and potential targets for biotechnological applications. Techniques include:

Mass Spectrometry: Identifies and quantifies proteins, helping to understand fungal metabolic processes and interactions with their environment. Mass spectrometry (MS) is a powerful analytical technique used to identify and quantify the chemical composition of fungal metabolites and other compounds. In mass spectrometry, a sample is ionized, and the resulting ions are separated based on their mass-to-charge ratios. This separation is achieved using a mass analyzer, and the intensity of the detected ions is used to determine the abundance of various compounds. For soil fungi, MS can provide detailed information on the metabolic profile, including the presence and concentration of secondary metabolites, enzymes, and other bioactive compounds. By analyzing these profiles, researchers can gain insights into fungal metabolic pathways, interactions with other organisms, and responses to environmental changes. MS is often used in conjunction with other techniques, such as liquid chromatography (LC-MS) or gas chromatography (GC-MS), to enhance the resolution and accuracy of the analysis. This technique is essential for studying fungal biochemistry, understanding their ecological roles, and exploring potential applications in biotechnology and environmental management.

4. Ecological and Functional Approaches

Ecological and functional approaches in the study of soil fungi focus on understanding their roles and interactions within ecosystems and their contributions to ecosystem processes. Ecological approaches involve examining fungal diversity, distribution, and community structure in different soil environments to understand how fungi influence and are influenced by their surroundings. This includes studying fungal interactions with plants, other microorganisms, and soil properties to assess their impact on soil health, nutrient cycling, and ecosystem stability. Functional approaches, on the other hand, analyze the specific ecological roles of fungi, such as their ability to decompose organic matter, fix nitrogen, or form symbiotic relationships with plants. By integrating these approaches, researchers can gain insights into how fungal communities function within ecosystems, how they adapt to environmental changes, and their overall contributions to ecosystem services. This holistic understanding helps in managing soil health, optimizing agricultural practices, and conserving biodiversity by highlighting the critical functions that fungi perform in maintaining ecosystem balance and resilience.

Soil Sampling and Processing: Collecting and processing soil samples from different environments helps in understanding fungal diversity and distribution. Soil sampling and processing are crucial steps in studying soil fungi, as they ensure representative and reliable data for analysis. The process begins with systematic soil sampling, where soil is collected from various locations within a study area to capture spatial variability. Samples are typically obtained using soil cores or augers and are carefully handled to prevent contamination. Once collected, soil samples are processed by air-drying or refrigerating to preserve microbial integrity and reduce microbial activity before analysis. The soil is then sieved to remove large debris and homogenized to ensure uniformity. For fungal analysis, soil is often mixed with sterile water or buffer to extract fungal spores and mycelium. The resulting slurry is then used for culturing fungi on selective media, or for molecular analyses such as DNA extraction. Proper soil processing is essential for accurate fungal identification, quantification, and understanding of fungal diversity and functions within the soil ecosystem. Key steps include:

Soil Extraction: Using techniques like soil dilution plating or soil washing to isolate fungi. Soil extraction is a fundamental procedure in soil microbiology and fungal studies, used to isolate and analyze microorganisms, including fungi, from soil samples. The process involves mixing soil with a suitable extraction medium, such as a sterile buffer or water, to dislodge and suspend soil particles and microbial cells. This mixture is then subjected to various techniques, such as shaking or homogenization, to ensure thorough extraction of fungi and other microorganisms. The resulting suspension can be filtered or centrifuged to separate fungal spores and mycelium from larger soil particles and debris. Soil extraction is crucial for subsequent analyses, including fungal culturing, molecular studies, and metabolic profiling. Proper extraction techniques are essential for obtaining representative samples and accurate data on fungal diversity, abundance, and activity, which are vital for understanding their roles in soil ecosystems and their interactions with other soil components.

Environmental Parameter Measurement: Assessing soil factors such as moisture, pH, and temperature to correlate with fungal distribution and activity. Environmental parameter measurement is crucial for understanding the impact of environmental conditions on soil fungi and their functions within ecosystems. This involves quantifying factors such as soil moisture, pH, temperature, and nutrient levels, which can significantly influence fungal growth, activity, and community structure. Techniques for measuring these parameters include using pH meters for soil acidity, thermometers or temperature probes for monitoring soil temperature, and moisture sensors or gravimetric methods for assessing water content. Nutrient levels can be determined through chemical assays or spectrometric methods. Accurate measurement of these parameters helps researchers correlate environmental conditions with fungal dynamics, revealing how fungi adapt to and influence their surroundings. By integrating environmental parameter measurements with fungal studies, scientists can better understand how soil fungi contribute to nutrient cycling, soil health, and ecosystem resilience, and how changes in environmental conditions may impact their ecological roles.

Stable Isotope Tracing: This technique tracks fungal roles in nutrient cycling by using isotopically labeled compounds (e.g., ^{15}N or ^{13}C) to trace the movement and transformation of nutrients through fungal processes. Stable isotope tracing is a sophisticated analytical technique used to track the movement and transformation of elements within soil fungi and their interactions with the environment. In this method, stable isotopes, such as carbon-13 (^{13}C) or nitrogen-15 (^{15}N), are introduced into the soil or directly into fungal systems, replacing their naturally occurring counterparts. These isotopes are then incorporated into fungal biomass and metabolic products, allowing researchers to trace their pathways and transformations. By analyzing the isotopic composition of fungal tissues or soil samples over time, scientists can gain insights into processes such as nutrient uptake, carbon sequestration, and nutrient cycling. Stable isotope tracing provides valuable information on how fungi assimilate nutrients from the soil, their role in food webs, and their interactions with plants and other microorganisms. This technique enhances our understanding of fungal ecology and contributes to studies on soil health, ecosystem functioning, and the impacts of environmental changes on nutrient dynamics.

Functional Assays: Assessing fungal activities, such as enzyme production (e.g., cellulases, laccases) or nutrient uptake, helps in characterizing their ecological roles. Functional assays are specialized techniques used to evaluate the specific activities and physiological functions of soil fungi, providing insights into their ecological roles and capabilities. These assays measure various fungal processes, such as enzyme production, nutrient uptake, and interactions with substrates. For example, enzyme activity assays involve adding substrates to fungal cultures and measuring the release of products to assess the activity of enzymes like cellulases, ligninases, or phosphatases, which are crucial for decomposing organic matter and cycling nutrients. Other functional assays might involve testing fungal ability to solubilize minerals or to form symbiotic relationships with plants. These assays help elucidate how fungi contribute to soil health, nutrient availability, and ecosystem processes, offering a deeper understanding of their functional roles in nutrient cycling, organic matter decomposition, and plant growth support. By evaluating these functions, researchers can better appreciate the ecological significance of soil fungi and their potential applications in agriculture and environmental management. Methods include:

Enzyme Activity Tests: Measuring the activity of specific enzymes involved in decomposition or nutrient cycling. Enzyme activity tests are essential for assessing the functional capabilities of soil fungi by measuring their ability to produce specific enzymes that catalyze biochemical reactions. In these tests, soil or fungal samples are exposed to substrates that are transformed by fungal enzymes, and the resulting products are quantified to determine enzyme activity. For example, assays may measure the activity of cellulases that break down cellulose into simpler sugars, or phosphatases that release phosphate from organic compounds. These tests can be conducted using colorimetric, fluorometric, or spectrophotometric methods, where changes in color or fluorescence indicate the extent of enzyme activity. By evaluating enzyme activity, researchers can gain insights into fungal roles in organic matter decomposition, nutrient cycling, and soil health. Enzyme activity tests help elucidate the functional contributions of fungi to soil ecosystems, their metabolic potential, and their interactions with other microorganisms and soil components.

Nutrient Uptake Studies: Evaluating how fungi contribute to nutrient absorption and transfer in soil. Nutrient uptake studies are vital for understanding how soil fungi acquire and utilize essential nutrients from their environment, which directly impacts their role in soil health and ecosystem functioning. These studies involve tracing the absorption and assimilation of nutrients such as nitrogen, phosphorus, and potassium by fungal mycelium. Techniques often used include radioisotope labeling, where specific isotopes of nutrients are introduced into the soil and tracked as they are taken up by fungi, and stable isotope analysis, which provides detailed insights into nutrient cycling and availability. Additionally, nutrient uptake studies can measure changes in fungal growth and metabolic activity in response to different nutrient levels or forms. This information helps to elucidate how fungi contribute to nutrient cycling, support plant growth through mycorrhizal associations, and affect soil fertility. By understanding the mechanisms and efficiency of nutrient uptake, researchers can better appreciate the ecological roles of soil fungi and their potential applications in sustainable agriculture and soil management.

Metabolomics: Profiling fungal metabolites provides insights into fungal physiology and their impact on soil processes. Metabolomics is a comprehensive analytical technique that examines the complete set of metabolites present in soil fungi, providing insights into their biochemical activity and ecological roles. By using methods such as mass spectrometry (MS) and nuclear magnetic resonance (NMR) spectroscopy, researchers can profile a wide range of metabolites, including small molecules, secondary metabolites, and metabolic intermediates. Metabolomics allows for the identification of metabolic pathways, the assessment of metabolic responses to environmental changes, and the discovery of fungal metabolites with potential ecological or industrial significance. For soil fungi, metabolomics can reveal how they interact with their environment, decompose organic matter, or contribute to nutrient cycling by analyzing changes in metabolite profiles in response to different soil conditions or treatments. This technique enhances our understanding of fungal physiology and metabolism, offering insights into their functional roles within soil ecosystems and their potential applications in biotechnology and environmental management. Techniques include:

Nuclear Magnetic Resonance (NMR): Analyzes metabolic profiles and identifies fungal metabolic products. Nuclear Magnetic Resonance (NMR) spectroscopy is a powerful analytical technique used to investigate the molecular structure, dynamics, and interactions of compounds, including those produced by soil fungi. NMR works by applying a magnetic field to a sample, causing nuclei of certain atoms (commonly hydrogen-1 or carbon-13) to resonate at specific frequencies. This resonance produces spectra that provide detailed information about the chemical environment and connectivity of atoms within molecules. In soil fungal studies, NMR can be used to analyze complex mixtures of metabolites, elucidate the structure of fungal secondary metabolites, and study metabolic changes in response to different environmental conditions. NMR's non-destructive nature and ability to provide quantitative data make it particularly valuable for characterizing fungal metabolic profiles, understanding biochemical pathways, and exploring interactions between fungi and their environment. This technique contributes to a deeper comprehension of fungal biology and their roles in soil ecosystems, offering insights into their ecological functions and potential applications in areas such as agriculture and biotechnology.

Liquid Chromatography-Mass Spectrometry (LC-MS): Detects and quantifies a wide range of fungal metabolites. Liquid Chromatography-Mass Spectrometry (LC-MS) is a sophisticated analytical technique that combines the separation power of liquid chromatography with the detailed identification capabilities of mass spectrometry, making it invaluable for studying soil fungi. In LC-MS, liquid chromatography first separates complex mixtures of metabolites or other compounds based on their chemical properties as they pass through a chromatographic column. Following separation, the compounds are introduced into the mass spectrometer, which measures their mass-to-charge ratios to identify and quantify them. This technique allows for the detailed profiling of fungal metabolites, including secondary metabolites, enzymes, and other bioactive compounds, providing insights into fungal metabolic pathways and ecological roles. LC-MS is particularly useful for detecting low-abundance compounds and complex mixtures, enabling researchers to explore the diversity and function of fungal metabolites in soil ecosystems. By analyzing metabolic changes and interactions, LC-MS helps to elucidate the biochemical activities of soil fungi and their contributions to nutrient cycling and soil health.

Combining these techniques provides a comprehensive understanding of soil fungi, from their identification and taxonomy to their ecological functions and contributions to soil health. Each method offers unique insights, and together they help build a detailed picture of the fungal diversity and dynamics within soil ecosystems.

8.2. Soil sampling and fungal culture techniques.

Soil sampling and fungal culture techniques are essential for studying soil fungi and understanding their roles within ecosystems. The process begins with careful soil sampling, where representative soil cores are collected from various depths and locations to capture the diversity of fungal populations. Using clean, sterile tools, researchers gather soil samples into sterile containers and transport them under controlled conditions to prevent contamination. Once in the laboratory, the soil is processed, often by air-drying or sieving, to prepare it for fungal isolation. Soil suspensions are then prepared through serial dilution, which reduces the concentration of fungi and facilitates the isolation of individual colonies. This is followed by plating the soil suspensions onto selective or general agar media, such as Potato Dextrose Agar (PDA) or Sabouraud Dextrose Agar (SDA), to support the growth of fungi. Incubation at optimal temperatures allows fungal colonies to develop, which are then observed for morphological characteristics like color, texture, and growth patterns. Further identification involves microscopic examination of the colonies to assess fungal structures such as spores and hyphae. To ensure purity and accurate identification, isolated fungi are sub-cultured onto fresh plates, and any contaminants are removed. Long-term preservation of fungal cultures is achieved through refrigeration, cryopreservation in liquid nitrogen, or lyophilization. Molecular techniques, including PCR and DNA sequencing, complement traditional methods by providing detailed genetic information that confirms species identification and reveals broader fungal diversity. These combined approaches enable comprehensive study of soil fungi, shedding light on their ecological roles, interactions, and contributions to soil health.

Soil Sampling

Soil sampling is a critical step in studying soil fungi, as it provides the material needed for isolating and identifying fungal species. Proper sampling techniques ensure that the samples are representative of the soil environment being studied.

1. Sampling Procedure:

- **Selection of Sampling Sites**: Choose locations based on the study objectives, such as different land uses, depths, or ecological zones. Ensure that the sites represent the variability of the area to get a comprehensive view of the fungal diversity.
- **Sampling Depth**: Depending on the study, collect soil samples from different depths, such as the top 0-10 cm (surface soil) and deeper layers (e.g., 10-30 cm). Different fungi may inhabit different soil depths.
- **Collection Tools**: Use clean, sterile tools like soil corers or augers to collect soil samples. Avoid contamination by ensuring that tools are free from residues and that samples are handled with clean gloves.
- **Sample Handling**: Collect soil samples in sterilized containers (e.g., plastic bags or glass jars) and transport them to the laboratory as soon as possible. Store samples at appropriate temperatures (e.g., 4°C) to preserve fungal viability until processing.

2. Sample Processing:

- **Soil Preparation**: Air-dry or sieve the soil to remove large debris and homogenize the sample. For specific studies, soil may be processed fresh to maintain the viability of certain fungi.
- **Soil Dilution**: To isolate fungi, prepare serial dilutions of the soil sample in sterile water or buffer solution. This reduces the concentration of fungi, making it easier to isolate individual colonies.

Fungal Culture Techniques

Culturing soil fungi allows researchers to isolate and identify fungal species based on their growth and morphological characteristics. Here's a detailed overview of common fungal culture techniques:

1. Isolation Methods:

- **Plating**: Diluted soil samples are spread onto agar plates to isolate fungi. Different types of agar media can be used, depending on the target fungi:

  Potato Dextrose Agar (PDA): Supports the growth of a wide range of fungi and is commonly used for general isolation.

  Sabouraud Dextrose Agar (SDA): Used for cultivating yeasts and molds.

  Selective Media: Media containing specific inhibitors or nutrients to select for certain fungal groups (e.g., Rose Bengal Agar for isolating fungi from soil).

- **Soil Dilution Plating**: Soil suspensions are spread on agar plates and incubated. Colonies that develop are then sub-cultured to obtain pure cultures.

2. Culturing Techniques:

- **Incubation**: Plates are incubated at suitable temperatures for fungal growth, typically between 20°C and 30°C. Incubation times can vary depending on the fungal species and their growth rates.
- **Observing Colony Morphology**: After incubation, fungal colonies are examined for characteristics such as color, texture, and growth pattern. These observations help in preliminary identification.
- **Microscopic Examination**: Isolated fungal colonies are examined under a microscope to observe structures such as spores, hyphae, and reproductive structures. This morphological information aids in species identification.

3. Sub-Culturing and Purification:

- **Sub-Culturing**: To obtain pure cultures, fungal colonies are transferred to fresh agar plates. This helps eliminate contaminants and ensures that only the target fungus is cultured.
- **Purification**: Techniques such as streaking for single colonies or using selective media can further purify cultures. Each isolated fungus can be transferred to a new plate to ensure purity.

4. Preservation of Fungal Cultures:

- **Storage Methods**: To maintain fungal cultures for long-term studies, they can be preserved using various methods:

  Refrigeration: Cultures can be stored at 4°C for short-term preservation.

Cryopreservation: Fungi can be preserved in liquid nitrogen at -196°C, ensuring long-term storage without loss of viability.

Lyophilization: Fungi can be freeze-dried and stored in a desiccated environment for long-term preservation.

5. Identification and Characterization:

- **Morphological Identification**: Detailed examination of colony morphology, spore structure, and other reproductive features under a microscope.
- **Molecular Techniques**: Molecular methods, such as PCR and DNA sequencing, can confirm fungal identification by analyzing genetic material. This helps in accurate species identification and understanding fungal diversity.

By combining soil sampling techniques with effective fungal culture methods, researchers can isolate, identify, and study soil fungi, providing valuable insights into their diversity, ecological roles, and interactions within soil ecosystems.

8.3. Advances in fungal genomics and bioinformatics in soil studies.

Advances in fungal genomics and bioinformatics have revolutionized soil studies, providing deep insights into fungal diversity, function, and ecology. These advancements have enabled researchers to explore the genetic and functional complexity of soil fungi with unprecedented precision and depth. Advances in fungal genomics and bioinformatics have significantly enhanced our understanding of soil fungi and their ecological roles. High-throughput sequencing technologies, such as Illumina and Oxford Nanopore, enable comprehensive profiling of fungal genomes and metagenomes, revealing detailed insights into fungal diversity and community composition in soil environments. These advancements allow for high-resolution genome assemblies and functional annotations, facilitating comparative genomics and the study of gene functions. Bioinformatics tools have further revolutionized fungal research by enabling metagenomic analyses that uncover the complex interactions within fungal communities. Databases like UNITE and FunGene, combined with computational models and network analysis tools, provide valuable resources for identifying species, predicting functions, and visualizing interactions. Together, these advancements offer a deeper understanding of the genetic and functional complexity of soil fungi, highlighting their critical roles in nutrient cycling, ecosystem health, and responses to environmental changes.

Advances in Fungal Genomics

High-Throughput Sequencing: High-throughput sequencing technologies, such as Illumina sequencing and Oxford Nanopore sequencing, have dramatically increased the ability to sequence fungal genomes and metagenomes. These technologies allow researchers to sequence entire fungal genomes or large portions of them, revealing detailed genetic information. Metagenomic sequencing, which involves sequencing the collective DNA from soil samples, provides a comprehensive overview of the fungal community present in a given environment, including those fungi that are difficult to culture.

Genome Assembly and Annotation: Advancements in genome assembly techniques, such as genome-wide sequencing and assembly algorithms, have improved the quality and completeness of fungal genome assemblies. Tools like SPAdes and SOAPdenovo enable the construction of high-quality genome sequences from complex datasets. Annotation tools, such as AUGUSTUS and MAKER, help in identifying genes, regulatory elements, and functional annotations, providing insights into the metabolic pathways, ecological functions, and potential interactions of fungi.

Comparative Genomics: Comparative genomics allows researchers to compare the genomes of different fungal species to identify conserved genes, evolutionary relationships, and functional differences. This approach helps in understanding how different fungi adapt to their environments, including their roles in nutrient cycling and interactions with plants. Tools like OrthoFinder and Clustal Omega facilitate the comparison of genomic data across species.

Functional Genomics: Functional genomics involves studying gene expression and function to understand how fungal genes contribute to various biological processes. Techniques such as RNA sequencing (RNA-Seq) provide a detailed view of the transcriptome, revealing which genes are active under different conditions. This helps in identifying genes involved in stress responses, nutrient acquisition, and interactions with other organisms. CRISPR/Cas9 and RNA interference (RNAi) are used to manipulate fungal genes and study their functions in detail.

Advances in Bioinformatics

Metagenomics and Community Analysis: Bioinformatics tools have advanced the analysis of metagenomic data, allowing researchers to study soil fungal communities without culturing. Tools like QIIME and Mothur facilitate the analysis of high-throughput sequencing data, enabling the identification and quantification of fungal species, as well as the assessment of community diversity and composition. Functional profiling tools, such as HUMAnN and PICRUSt, predict the functional potential of microbial communities based on their genomic data.

Database Development and Utilization: The development of comprehensive fungal databases, such as UNITE and FunGene, provides valuable resources for identifying fungal species and understanding their ecological roles. These databases include curated sequences, taxonomic classifications, and functional annotations. Bioinformatics tools that integrate these databases, like BLAST and DIAMOND, enable researchers to match sequencing data to known fungal sequences and predict functions.

Systems Biology and Network Analysis: Systems biology approaches use bioinformatics to integrate genomic, transcriptomic, and proteomic data to understand complex fungal systems. Network analysis tools, such as Cytoscape, allow researchers to visualize and analyze interactions between genes, proteins, and metabolic pathways. This helps in understanding how fungal networks function in nutrient cycling, disease resistance, and other ecological processes.

Computational Modeling: Computational models simulate fungal growth, interactions, and ecological functions based on genomic and environmental data. Models like FUNGuild and Ecological Network Analysis help predict how fungal communities respond to environmental changes and manage ecosystems. These models integrate large datasets and simulate scenarios to understand the potential impacts of factors like climate change or land use on fungal populations.

Data Integration and Visualization: Bioinformatics tools for data integration and visualization, such as R and Python libraries (e.g., ggplot2, Pandas), facilitate the analysis of complex datasets and the presentation of results. These tools help researchers interpret large volumes of genomic and ecological data, identify trends, and communicate findings effectively.

Together, these advances in fungal genomics and bioinformatics enhance our understanding of soil fungi, their diversity, functions, and roles in ecosystems. They enable researchers to unravel the genetic and functional complexities of fungal communities, leading to better insights into their ecological significance and potential applications in agriculture, conservation, and environmental management.

8.4.Conclusion

The methods of studying soil fungi, encompassing soil sampling, fungal culture techniques, and advanced genomic and bioinformatics approaches, offer a comprehensive framework for understanding the intricate roles of fungi in soil ecosystems. Soil sampling provides the foundational material for research, with careful techniques ensuring that samples are representative and free from contamination. The subsequent fungal culture techniques enable the isolation and identification of fungal species, allowing researchers to observe their growth, morphology, and interactions in controlled conditions. These traditional methods are complemented by modern advances in fungal genomics and bioinformatics, which offer powerful tools for analyzing fungal diversity, function, and ecology with high resolution. High-throughput sequencing technologies and genome assembly techniques have revolutionized our ability to explore fungal genomes, while functional genomics and comparative genomics provide insights into the roles and evolutionary relationships of different fungi.

The integration of bioinformatics tools further enhances the study of soil fungi, allowing for detailed metagenomic analysis, functional profiling, and systems biology approaches. These tools facilitate the examination of complex fungal communities, prediction of ecological functions, and visualization of data, thereby advancing our understanding of how fungi contribute to soil health and ecosystem dynamics. Together, these methodologies not only elucidate the diverse and dynamic nature of soil fungal communities but also highlight their crucial roles in nutrient cycling, plant interactions, and ecosystem stability. As research continues to evolve, the synergy between traditional techniques and cutting-edge technologies promises to deepen our knowledge of soil fungi and their impact on environmental sustainability.

Chapter 9

Challenges and Future Directions in Soil Fungal Research

Soil fungi research encompasses the study of fungal species that inhabit soil environments, focusing on their diversity, ecology, and functional roles within ecosystems. Soil fungi play a pivotal role in nutrient cycling, organic matter decomposition, and soil structure formation, making them integral to soil health and ecosystem functioning. Research in this field often involves exploring fungal biodiversity through both traditional methods, such as morphological and cultural techniques, and modern molecular approaches, including DNA sequencing and metagenomics. This research aims to identify and classify various fungal species, understand their interactions with other soil microorganisms and plants, and assess their contributions to ecosystem processes. However, the vast diversity and complexity of soil fungi, along with challenges in culturing and identifying many species, present significant obstacles to comprehensive understanding. Advancements in soil fungi research are driven by technological innovations and interdisciplinary approaches. The integration of high-throughput sequencing technologies, bioinformatics tools, and multi-omics analyses has revolutionized our ability to characterize fungal communities at unprecedented scales and resolutions. These advancements allow researchers to delve deeper into the functional roles of soil fungi, including their responses to environmental changes, land management practices, and climate stressors. Future research directions emphasize the need for a more nuanced understanding of fungal-plant interactions, the impact of fungal biodiversity on ecosystem services, and the potential applications of soil fungi in sustainable agriculture and environmental remediation. Collaborative efforts and continued methodological advancements are essential for addressing the complexities of soil fungal ecosystems and harnessing their full potential in addressing global challenges.

Soil fungal research faces several significant challenges that impact our understanding of these crucial microorganisms. One major challenge is the complexity and diversity of soil fungal communities, which makes comprehensive characterization and monitoring difficult. Soil is a highly heterogeneous environment with varying physical, chemical, and biological conditions that can influence fungal distribution and function. This complexity, combined with the vast number of fungal species, many of which are not yet described or understood, complicates efforts to identify and study individual fungi accurately. Additionally, traditional culturing methods often fail to capture the full diversity of soil fungi, as many species are not easily cultured in laboratory settings. The reliance on molecular techniques like metagenomics and sequencing can generate large datasets, but interpreting this data and linking it to specific ecological functions remains challenging. Looking towards the future, advancements in technology and methodology offer promising directions for overcoming these challenges. Integrating multi-omics approaches, including genomics, transcriptomics, proteomics, and metabolomics, can provide a more holistic view of fungal communities and their functions. Improved bioinformatics tools and databases are essential for managing and analyzing complex data, enabling more accurate identification and functional annotation of fungal species. Additionally, there is a growing emphasis on exploring the functional roles of soil fungi in various ecosystems, including their responses to climate change, land management practices, and environmental stressors. Collaborative research efforts and interdisciplinary approaches are crucial for advancing our understanding of soil fungal ecology and harnessing their potential for applications in agriculture, environmental management, and biotechnology.

9.1. Gaps in knowledge and current challenges in soil fungal ecology.

Despite significant progress in soil fungal ecology, several critical gaps in knowledge and challenges persist. One major issue is the incomplete understanding of fungal diversity and their ecological roles, particularly in less-studied regions or extreme environments. Many soil fungi remain unidentified due to limitations in culturing techniques and taxonomic resources. Additionally, the complexity of soil ecosystems, characterized by dynamic interactions between fungi, plants, and other microorganisms, complicates efforts to elucidate specific functional roles and relationships. The impact of environmental changes, such as climate change and land-use alterations, on fungal communities and their functions is also not fully understood, particularly concerning their adaptive responses and contributions to ecosystem resilience. Addressing these gaps requires a concerted effort to develop and apply advanced methodologies, such as high-throughput sequencing and integrative multi-omics approaches, alongside increased focus on field studies and long-term ecological monitoring.

Gaps in knowledge in soil fungal ecology

Gaps in soil fungal ecology persist in several critical areas, primarily due to the vast and complex nature of fungal diversity and their interactions within soil ecosystems. Despite advances in molecular and sequencing technologies, many soil fungi remain unidentified or inadequately characterized, leaving significant gaps in our understanding of their taxonomy and ecological roles. The specific functional contributions of various fungal species to soil health and nutrient cycling are not fully elucidated, particularly in the context of complex microbial interactions and responses to environmental stressors. Additionally, the impacts of global environmental changes, such as climate shifts and land-use alterations, on soil fungal communities and their functions remain poorly understood. Addressing these gaps necessitates enhanced methodological approaches, including comprehensive field surveys, integrative multi-omics techniques, and sophisticated ecological modeling to provide a more complete picture of soil fungal dynamics and their roles in ecosystem processes. Gaps in knowledge in soil fungal ecology include several key areas where understanding is still limited or evolving:

Fungal Diversity and Classification: Despite advances in molecular techniques, many soil fungal species remain undiscovered or poorly characterized. The vast diversity of soil fungi and the challenges associated with culturing and identifying them contribute to incomplete taxonomic databases. There is a need for comprehensive surveys and improved taxonomic frameworks to better capture and classify the wide range of fungal species present in soil environments.

Functional Roles and Interactions: While the ecological roles of soil fungi in nutrient cycling, organic matter decomposition, and soil structure formation are recognized, the specific functional contributions of many fungal species are not well understood. Research often focuses on a limited number of model species, leaving gaps in knowledge about how diverse fungal communities contribute to ecosystem processes and interact with other soil microorganisms and plants.

Impact of Environmental Changes: The effects of environmental changes, such as climate change, land use alterations, and pollution, on soil fungal communities and their functions are not fully understood. There is a need to study how these changes impact fungal diversity, distribution, and ecological roles, as well as how fungi adapt to and influence ecosystem responses to such stressors.

Fungal-Plant Interactions: While mycorrhizal fungi and their role in plant nutrient uptake are well-documented, other types of fungal-plant interactions, such as endophytic and pathogenic relationships, require further investigation. Understanding these interactions is crucial for managing plant health and optimizing agricultural practices.

Quantitative Data and Modeling: There is a need for more quantitative data on soil fungal biomass, activity, and interactions within soil ecosystems. Additionally, developing and refining models that predict fungal community dynamics and their impacts on soil health and ecosystem functions will enhance our ability to manage and conserve soil ecosystems effectively.

Addressing these gaps requires a multidisciplinary approach, combining field studies, advanced molecular techniques, and ecological modeling to gain a more comprehensive understanding of soil fungi and their roles in various environments.

Current challenges in soil fungal ecology

Current challenges in soil fungal ecology include several significant issues that impede a comprehensive understanding of fungal communities and their roles in ecosystems. One major challenge is the immense diversity of soil fungi, which complicates efforts to identify and classify species accurately. Many fungi are difficult to culture and study using traditional methods, leading to incomplete knowledge about their taxonomy and functional roles. Additionally, the complex interactions between soil fungi, plants, and other microorganisms are not fully understood, particularly in terms of how these relationships influence soil health and ecosystem processes. Environmental changes, such as climate change and land-use modifications, further exacerbate these challenges by altering fungal communities and their functions in unpredictable ways. There is also a need for better quantitative data and predictive models to understand how changes in fungal populations affect soil dynamics and ecosystem services. Addressing these challenges requires a multidisciplinary approach, combining advanced molecular techniques, field studies, and ecological modeling to provide a more integrated and actionable understanding of soil fungal ecology. Current challenges in soil fungal ecology are multifaceted and impact the depth and breadth of our understanding of these crucial organisms:

Diversity and Identification: Soil fungi exhibit immense diversity, with a large number of species remaining undescribed or poorly understood. Traditional culturing techniques often fail to capture the full spectrum of soil fungi, as many are difficult or impossible to grow in laboratory conditions. This limitation is compounded by the complexity of soil environments, where fungi interact with a myriad of other microorganisms and plants, making identification and classification challenging. Advanced molecular techniques, such as high-throughput sequencing, have made significant strides, but accurately linking sequence data to specific fungal species and their ecological roles remains difficult.

Functional Understanding: While the roles of soil fungi in nutrient cycling, organic matter decomposition, and soil structure formation are recognized, the specific functions of many fungal species and their interactions within soil ecosystems are not well understood. The functional diversity of soil fungi is influenced by factors such as microbial interactions, soil conditions, and plant relationships, which are complex and not always well-characterized. There is a need for more detailed studies that link fungal diversity to functional outcomes, including how different species contribute to soil health and ecosystem processes.

Impact of Environmental Changes: Climate change, land-use alterations, and pollution have significant impacts on soil fungal communities, but the precise effects and mechanisms are not fully understood. These environmental changes can alter fungal diversity, distribution, and activity, which in turn affects soil health and ecosystem functions. Research must address how soil fungi adapt to changing conditions and how their altered roles impact soil and plant interactions. There is also a need to understand how these changes influence fungal contributions to global biogeochemical cycles, such as carbon and nitrogen cycling.

Quantitative Data and Modeling: There is a lack of comprehensive quantitative data on soil fungal biomass, activity, and interactions. This data is essential for developing accurate models that predict fungal community dynamics and their impacts on soil health. Current models often rely on incomplete data and may not fully capture the complexities of fungal behavior and interactions. Improved data collection methods and more sophisticated modeling approaches are needed to enhance our ability to manage and conserve soil ecosystems effectively.

Integration of Multidisciplinary Approaches: Soil fungal ecology benefits from an integrative approach that combines molecular biology, field ecology, and computational modeling. However, coordinating these diverse methods can be challenging. Effective integration requires collaboration across disciplines, as well as the development of new tools and methodologies that can bridge gaps between different types of data and provide a more holistic understanding of soil fungal ecology.

Addressing these challenges involves a concerted effort to advance technological and methodological approaches, enhance interdisciplinary collaboration, and focus on the ecological and functional relevance of soil fungi in various environments.

9.2. Emerging technologies in fungal research.

Emerging technologies in fungal research are significantly advancing our understanding of fungal biology, ecology, and applications. High-throughput sequencing technologies, including next-generation sequencing (NGS) and metagenomics, enable comprehensive analysis of fungal communities and their genetic diversity, offering insights into species identification and functional potential. Metabolomics and proteomics provide detailed profiles of fungal metabolites and proteins, revealing their roles in biochemical processes and interactions with their environment. Advances in bioinformatics and computational tools facilitate the integration and interpretation of complex omics data, enhancing our ability to model fungal functions and interactions. Additionally, developments in imaging techniques, such as high-resolution microscopy and imaging mass spectrometry, allow for the visualization of fungal structures and activities at unprecedented detail. These technologies are transforming fungal research by enabling more precise and expansive exploration of fungal ecosystems, their roles in various environments, and their potential applications in agriculture, medicine, and biotechnology.

Emerging technologies in fungal research are revolutionizing the field by enhancing our ability to explore and understand fungal biology, ecology, and applications. Key advancements include:

High-Throughput Sequencing (HTS): Technologies such as Next-Generation Sequencing (NGS) and third-generation sequencing (e.g., Oxford Nanopore, PacBio) are transforming fungal research by providing deep insights into fungal diversity and genome complexity. HTS allows for the sequencing of entire fungal communities from environmental samples, enabling researchers to identify and quantify species that are difficult to culture. These technologies also facilitate the study of fungal genomics, including gene function, genetic variation, and evolutionary relationships.

Metagenomics: This approach involves the study of genetic material recovered directly from environmental samples, bypassing the need for cultivation. Metagenomics allows researchers to analyze the entire microbial community in soil, including fungi, and assess their functional potential and interactions within their ecosystems. This technology provides a comprehensive view of fungal diversity and functional capabilities, helping to elucidate the roles of fungi in nutrient cycling and other ecological processes.

Metabolomics and Proteomics: Metabolomics involves the analysis of metabolites produced by fungi, offering insights into their biochemical activities and interactions with the environment. Proteomics, on the other hand, focuses on the study of fungal proteins and their functions. Together, these technologies provide a detailed understanding of fungal metabolic pathways, enzyme functions, and responses to environmental changes.

Bioinformatics and Computational Tools: Advanced bioinformatics tools are essential for analyzing and interpreting the vast amounts of data generated by sequencing and omics technologies. These tools help in assembling fungal genomes, annotating genes, and identifying functional pathways. Computational models and network analysis are used to understand fungal interactions, predict functional roles, and simulate ecological dynamics.

Imaging Techniques: Innovations in imaging technologies, such as high-resolution microscopy (e.g., confocal microscopy, electron microscopy) and imaging mass spectrometry, allow for the detailed visualization of fungal structures and processes. These techniques enable researchers to study fungal growth, spore development, and interactions with plants and other microorganisms at a microscopic level.

Synthetic Biology: This field applies engineering principles to fungal biology, enabling the modification of fungal genomes for specific purposes. Synthetic biology approaches are used to create genetically engineered fungi with enhanced capabilities for applications in agriculture, biotechnology, and medicine, such as improved production of pharmaceuticals or biocontrol agents.

Functional Genomics: Techniques like CRISPR-Cas9 gene editing allow for precise modifications of fungal genomes to study gene function and validate targets for biotechnological applications. Functional genomics enables researchers to investigate the roles of specific genes in fungal development, metabolism, and interactions.

Environmental DNA (eDNA) Analysis: eDNA techniques involve detecting and analyzing DNA fragments present in environmental samples such as soil, water, or air. This method provides insights into the presence and abundance of fungal species in various environments, including those that are difficult to sample directly.

These emerging technologies are expanding the frontiers of fungal research by providing new tools and methodologies to explore fungal diversity, functions, and applications. They are crucial for advancing our understanding of fungal roles in ecosystems and leveraging fungi for practical applications in agriculture, industry, and medicine.

9.3.Future prospects for fungi in biotechnology, medicine, and environmental science.

The future prospects for fungi in biotechnology, medicine, and environmental science are exceptionally promising due to their versatile and dynamic roles. In biotechnology, fungi are increasingly harnessed for their ability to produce valuable metabolites, enzymes, and bioactive compounds, with applications ranging from sustainable agriculture—such as biopesticides and biofertilizers—to industrial processes like bioremediation and biofuel production. In medicine, fungi offer potential as sources of novel pharmaceuticals, including antibiotics, immunosuppressants, and anticancer agents, with ongoing research aimed at uncovering new therapeutic compounds and understanding their mechanisms of action. Additionally, fungi's natural abilities in environmental science are being leveraged for ecological restoration, soil health improvement, and the mitigation of climate change impacts, such as carbon sequestration and pollutant degradation. Advances in genomics, synthetic biology, and molecular techniques are poised to enhance our ability to exploit fungal resources and optimize their applications, thereby driving innovation and addressing critical challenges in health, industry, and environmental sustainability.

The future prospects for fungi in biotechnology

The future prospects for fungi in biotechnology are expansive, reflecting their diverse applications and potential for innovation across various sectors. Here are some detailed areas where fungi are poised to make significant impacts:

Enzyme Production: Fungi are renowned for their ability to produce a wide range of enzymes with industrial applications. These enzymes, such as cellulases, ligninases, and proteases, are essential for processes in the pulp and paper industry, biofuel production, and food processing. Advances in fungal genomics and protein engineering are likely to enhance enzyme efficiency and specificity, leading to more sustainable and cost-effective industrial processes.

Pharmaceutical Development: Fungi are a rich source of bioactive compounds with therapeutic potential. Historically, fungi have provided antibiotics like penicillin and antifungals like cyclosporine. Future research aims to discover new drugs and therapeutic agents from fungal metabolites, focusing on treating diseases such as cancer, diabetes, and neurodegenerative disorders. Advances in synthetic biology and high-throughput screening are expected to accelerate the discovery and development of novel pharmaceuticals.

Biopesticides and Biofertilizers: The use of fungi as biopesticides and biofertilizers offers an eco-friendly alternative to chemical pesticides and fertilizers. Mycorrhizal fungi, for example, can enhance plant nutrient uptake and improve soil health, while entomopathogenic fungi can target agricultural pests. Ongoing research aims to optimize these applications, enhance the efficacy of fungal formulations, and develop new fungal-based products that promote sustainable agriculture.

Bioremediation: Fungi play a critical role in the bioremediation of contaminated environments, including soils and water bodies. Their ability to degrade pollutants, such as petroleum hydrocarbons, pesticides, and heavy metals, makes them valuable for environmental cleanup. Future advancements in fungal genetics and metabolic engineering are expected to improve the efficiency of bioremediation processes, enabling fungi to tackle a broader range of contaminants and operate under diverse environmental conditions.

Biomaterials and Biopolymers: Fungi can produce a variety of biomaterials and biopolymers, such as chitosan and fungal mycelium, which have applications in packaging, construction, and medical devices. Mycelium-based materials, in particular, are being explored for their potential as sustainable alternatives to synthetic materials. Research into optimizing the growth conditions and properties of fungal biomaterials is expected to lead to innovative products with enhanced performance and environmental benefits.

Synthetic Biology and Genetic Engineering: Advances in synthetic biology and genetic engineering are expanding the potential applications of fungi in biotechnology. By manipulating fungal genomes, researchers can engineer fungi to produce custom-designed metabolites, enzymes, or materials. Techniques such as CRISPR-Cas9 and other gene-editing tools are facilitating the development of fungal strains with tailored properties for specific industrial applications.

Biofuel Production: Fungi have potential in the production of biofuels through the fermentation of lignocellulosic biomass. Fungal enzymes that break down complex plant materials can be utilized in bioethanol production, contributing to the development of renewable energy sources. Future research aims to improve the efficiency of fungal-based biofuel processes and reduce production costs.

Food and Beverage Industry: Fungi are integral to the production of various food and beverage products, such as bread, beer, and cheese. Future prospects include the development of new fungal strains with enhanced flavor profiles, improved fermentation processes, and potential health benefits. Innovations in fungal fermentation technology may also lead to the creation of novel food products and functional ingredients.

The future of fungi in biotechnology is marked by ongoing research and technological advancements that promise to unlock new applications and enhance existing processes. Fungi's versatility and adaptability position them as key players in addressing global challenges in health, industry, and environmental sustainability.

Future prospects for fungi in medicine

The future prospects for fungi in medicine are exceptionally promising due to their unique biochemical properties and diverse metabolic capabilities. Here are several key areas where fungi are expected to play a significant role:

Drug Discovery and Development: Fungi are renowned for their production of bioactive compounds, many of which have therapeutic potential. The exploration of fungal biodiversity, including previously unexplored or under-studied species, holds promise for discovering novel drugs. Historically, fungal-derived antibiotics like penicillin and cephalosporins have revolutionized medicine. Future research is likely to uncover new classes of antibiotics, antifungals, and anticancer agents from fungi, particularly as we continue to identify and characterize diverse fungal metabolites.

Antibiotic Resistance: The increasing problem of antibiotic resistance has heightened the need for new antibiotics. Fungi, with their diverse secondary metabolites, offer a reservoir of potential new drugs that can address resistant strains of bacteria and fungi. Advanced screening methods and synthetic biology could facilitate the development of novel antifungal and antibacterial agents derived from fungi.

Immunotherapy: Fungi produce compounds that can modulate the immune system. These include immunosuppressants like cyclosporine, which is used in organ transplantation. Future research could expand on this by discovering additional fungal compounds with potential as immunomodulatory agents for treating autoimmune diseases or improving the efficacy of cancer immunotherapy.

Probiotics and Prebiotics: Fungal probiotics, or beneficial fungi, have the potential to support gut health and modulate the microbiome. Research into fungal strains that can produce prebiotics or exert beneficial effects on the gut microbiota may lead to new treatments for digestive disorders and broader applications in health maintenance.

Vaccine Development: Fungi can also contribute to vaccine development, either as sources of novel antigens or through the use of fungal expression systems to produce vaccine components. Their ability to produce complex proteins and carbohydrates can be harnessed to develop vaccines against various pathogens.

Metabolic Engineering: Advances in metabolic engineering and synthetic biology enable the modification of fungal strains to produce high-value pharmaceutical compounds more efficiently. By engineering fungi to express specific genes or produce particular metabolites, researchers can optimize the production of drugs and potentially reduce costs associated with pharmaceutical manufacturing.

Biomaterials: Fungi produce chitin and chitosan, which have potential applications in medical materials, including wound dressings and tissue engineering. Research into fungal-based biomaterials could lead to innovative solutions in regenerative medicine and surgical applications.

Diagnostics: Fungal biomarkers and metabolites can be used in diagnostic tests to detect diseases or monitor therapeutic responses. Improved analytical techniques could enhance the sensitivity and specificity of these tests, leading to better disease management.

Overall, the diverse metabolic capabilities of fungi and their ability to produce complex organic compounds position them as valuable resources for future medical advancements. Continued research and technological developments are likely to unlock new therapeutic potentials and address pressing health challenges.

Future prospects for fungi in environmental science

The future prospects for fungi in environmental science are robust and multifaceted, driven by their unique ecological roles and biochemical capabilities. Here's an in-depth look at how fungi might shape environmental science and sustainability in the coming years:

Bioremediation: Fungi are adept at breaking down organic pollutants and are increasingly recognized for their potential in bioremediation. They can metabolize complex organic compounds, such as those found in oil spills, pesticides, and heavy metals, converting them into less harmful substances. Future research is likely to focus on identifying and engineering fungal strains with enhanced capabilities for degrading a wider range of pollutants. Advances in genetic engineering and synthetic biology could create custom fungi designed to address specific contamination challenges in various environments.

Soil Health and Ecosystem Restoration: Fungi play crucial roles in maintaining soil health by contributing to nutrient cycling, organic matter decomposition, and soil structure formation. In ecosystem restoration, fungi can be used to rehabilitate degraded soils, enhance soil fertility, and support plant re-establishment. Mycorrhizal fungi, for instance, form symbiotic relationships with plants to improve nutrient uptake, especially in nutrient-poor soils. Future prospects include the development of fungal-based soil amendments and restoration techniques that can improve soil health and support sustainable land management practices.

Climate Change Mitigation: Fungi contribute significantly to carbon sequestration by decomposing organic matter and incorporating carbon into stable soil forms. Their role in the carbon cycle is critical in mitigating climate change. Future research could focus on understanding how different fungal species contribute to carbon storage and how to enhance their activity to increase carbon sequestration. This includes exploring fungal interactions with other soil microorganisms and plants to optimize their role in global carbon cycling.

Biofungicides and Biopesticides: Fungi have potential as biofungicides and biopesticides to control plant diseases and pests in agriculture. They offer an eco-friendly alternative to chemical pesticides, reducing environmental contamination and promoting sustainable agricultural practices. The development of new fungal biocontrol agents that target a broader range of pests and diseases, and the refinement of application techniques, will be key areas of future research.

Biofuel Production: Fungi are involved in the production of biofuels through the decomposition of lignocellulosic biomass. Their enzymatic systems can break down complex plant materials into fermentable sugars, which can then be converted into bioethanol or other biofuels. Advances in fungal genomics and metabolic engineering may enhance the efficiency of this process, making fungal-based biofuel production more viable and sustainable.

Environmental Monitoring and Biosensors: Fungi can be used as biosensors for environmental monitoring due to their sensitivity to changes in environmental conditions. They can detect pollutants, soil moisture levels, and other environmental parameters. Future developments could involve engineering fungi to produce measurable signals or metabolites in response to specific environmental changes, improving monitoring and early warning systems for environmental management.

Symbiosis and Plant-Microbe Interactions: Understanding and harnessing the symbiotic relationships between fungi and plants can improve ecosystem resilience and productivity. Future research may focus on optimizing these interactions to enhance plant growth, particularly in challenging environments such as arid or contaminated soils. This includes exploring the potential of fungal inoculants to improve plant stress tolerance and ecosystem stability.

Fungal Biotechnology: The application of fungal biotechnology in environmental science involves leveraging fungal capabilities for waste management, resource recovery, and sustainable materials production. This includes developing fungal-based processes for recycling industrial by-products and creating biodegradable materials from fungal biomass.

Overall, the future of fungi in environmental science is characterized by their potential to address a range of environmental challenges through biotechnological innovations and sustainable practices. Continued research and technological advancements will likely expand their applications and enhance their roles in promoting environmental health and sustainability.

9.4.Conclusion

The challenges and future directions in soil fungal research highlight both the complexity of fungal ecosystems and the potential for significant advancements in our understanding and application of these organisms. One of the primary challenges is the immense diversity of soil fungi, which complicates efforts to catalog and understand their roles in ecosystems. The vast number of fungal species, many of which remain undiscovered or poorly understood, presents a significant obstacle to comprehensive research. Additionally, the interactions between soil fungi and other microorganisms, plants, and environmental factors are highly intricate and not yet fully elucidated. To overcome these challenges, researchers must develop and refine high-throughput techniques for fungal identification, functional analysis, and ecosystem modeling. Advances in genomic, metagenomic, and bioinformatics tools offer promising avenues for unraveling the complex interactions and functional roles of soil fungi, but these technologies must be adapted and scaled to handle the vast diversity of fungal species.

Looking to the future, the integration of emerging technologies and interdisciplinary approaches will be crucial for advancing soil fungal research. The development of novel molecular and imaging techniques, coupled with improvements in data analytics and computational modeling, will enable a more detailed understanding of fungal functions and their contributions to ecosystem processes. Additionally, addressing the impacts of global challenges such as climate change and land use change on soil fungal communities will require innovative research approaches that incorporate environmental variables and stressors. Future research must also focus on translating fundamental fungal science into practical applications, such as enhancing soil health, improving agricultural sustainability, and addressing environmental contamination. By addressing these challenges and pursuing these future directions, soil fungal research can contribute to broader ecological and environmental goals, ultimately leading to more sustainable and resilient ecosystems.

Glossary of Key Terms

Agaricus bisporus: Commonly known as the white button mushroom, this species is widely cultivated and used as a food source. It is a type of Basidiomycota and plays a role in organic matter decomposition.

Ascomycota: A phylum of fungi characterized by the formation of spores in a sac-like structure called an ascus. This group includes important fungi such as yeast (Saccharomyces cerevisiae) and morels (Morchella spp.).

Basidiomycota: A phylum of fungi known for producing spores on basidia, which are specialized structures. This group includes mushrooms, rusts, and smuts. They play a significant role in nutrient cycling and decomposition.

Biocontrol Agents: Organisms used to control pests and diseases in agriculture. Fungi can act as biocontrol agents by competing with or parasitizing harmful microorganisms or pests.

Bioremediation: The use of organisms, including fungi, to remove or neutralize contaminants from the environment, such as in soil or water pollution.

Chytridiomycota: A phylum of fungi characterized by their motile zoospores with a single flagellum. They are primarily aquatic and include species that are important in decomposing organic matter.

Decomposer: An organism that breaks down dead organic material, recycling nutrients back into the ecosystem. Many fungi, including those in the Basidiomycota and Ascomycota, function as decomposers.

Ectomycorrhizae: A type of mycorrhizal association where fungi form a sheath around the roots of plants and extend into the soil, enhancing nutrient uptake, particularly phosphorus.

Endophytes: Fungi that live within plant tissues without causing disease. They can enhance plant growth, stress tolerance, and resistance to pathogens.

Fungal Genomics: The study of the complete genetic material of fungi. Advances in fungal genomics help in understanding the genetic basis of fungal functions, interactions, and adaptations.

Glomeromycota: A phylum of fungi that forms arbuscular mycorrhizae with plants. These fungi are crucial for nutrient uptake, particularly phosphorus, and play a significant role in soil health.

Hyphal Structure: The network of filamentous structures that make up the body of a fungus. Hyphae grow and spread through substrates, aiding in nutrient absorption and decomposition.

Lignocellulosic Biomass: Plant material composed of lignin and cellulose. Fungi can decompose lignocellulosic biomass, playing a key role in nutrient cycling and biofuel production

Metabolomics: The comprehensive analysis of metabolites in a biological sample. This technique helps in understanding the metabolic processes and functions of fungi.

Mycorrhizae: Symbiotic associations between fungi and plant roots. Mycorrhizae enhance nutrient uptake for plants and improve soil health.

Mycelium: The vegetative part of a fungus, consisting of a mass of branching hyphae. Mycelium is involved in nutrient absorption and decomposition.

Nutrient Cycling: The process by which nutrients are recycled in the ecosystem. Fungi play a vital role in decomposing organic matter and facilitating nutrient availability.

Polymerase Chain Reaction (PCR): A molecular technique used to amplify specific DNA sequences. PCR is widely used in fungal identification and genetic analysis.

Proteomics: The large-scale study of proteins, their functions, and interactions. In fungal research, proteomics helps in understanding the protein expression and function of fungi.

Spore Morphology: The study of the size, shape, and structure of fungal spores. Spore morphology is crucial for identifying fungal species and understanding their reproductive strategies.

Symbiotic Relationships: Interactions between different species that benefit at least one of the participants. Mycorrhizal associations between fungi and plants are a key example of such relationships.

Taxonomy: The classification of organisms into hierarchical categories based on shared characteristics. Fungal taxonomy involves grouping fungi into phyla, classes, orders, families, genera, and species.

Yeast: A type of single-celled fungus that reproduces asexually by budding or fission. Yeasts are used in baking, brewing, and as model organisms in genetic studies.

Zygomycota: A phylum of fungi characterized by the production of zygospores during sexual reproduction. This group includes fungi that are important in decomposition and food spoilage.

Fungal Bioinformatics: The application of computational tools and techniques to analyze fungal genetic data. Bioinformatics helps in understanding fungal diversity, function, and evolution.

Functional Gene Analysis: The study of genes to determine their roles and contributions to fungal functions. This analysis helps in understanding how fungi interact with their environment and other organisms.

Fungal Taxonomy: The science of classifying and naming fungi based on their evolutionary relationships and morphological characteristics.

References

1. Agarwal, P. K., & Kumar, R. (2011). "Fungal Biodiversity in Indian Soils: An Overview." *Indian Journal of Mycology*, 24(1), 12-21.

2. Agarwal, P., & Singh, M. (2013). "Soil Fungi of Indian Grasslands: Diversity and Ecological Roles." *Journal of Environmental Biology*, 34(2), 189-196.

3. Ainsworth, G. C., & Bisby, G. R. (1995). *Ainsworth & Bisby's Dictionary of the Fungi*. CAB International.

4. Ainsworth, G. C., & Bisby, G. R. (2004). *Dictionary of the Fungi*. CABI Publishing.

5. Alexander, M. (2000). *Biodegradation and Bioremediation*. Academic Press.

6. Allan, R. S., et al. (2008). "The role of fungal diversity in soil health." *Fungal Ecology*, 1(1), 1-10.

7. Anderson, J. M., & Cairns, W. J. (1982). *The Role of Soil Fungi in Soil Formation*. Academic Press.

8. Arora, D. (1986). *Mushrooms Demystified*. Ten Speed Press.

9. Artursson, V., et al. (2006). "Soil fungal diversity and function in relation to land use and management practices." *Soil Biology and Biochemistry*, 38(5), 1314-1324.

10. Baldrian, P. (2008). "Soil microorganisms and their roles in nutrient cycling." *Fungal Diversity*, 33, 37-48.

11. Baldrian, P. (2017). "Fungal community in the forest soil." *In: Soil Microbiology, Ecology, and Biochemistry*. CRC Press.

12. Bénédicte, C., et al. (2003). "The role of fungi in ecosystem functioning." *Ecosystems*, 6(5), 451-464.

13. Bertin, C., et al. (2009). "Soil fungi and bioremediation." *Applied Soil Ecology*, 41(3), 177-189.

14. Bhargava, P., & Sharma, A. (2014). "Mycorrhizal Associations in Indian Forest Soils." *Mycological Research*, 118(5), 281-290.

15. Blanke, M., et al. (2007). "Fungal communities in soil ecosystems." *Soil Biology and Biochemistry*, 39(12), 3064-3072.

16. Boddy, L. (2000). "Fungi as decomposers." *In: The Fungal Community*. CRC Press.

17. Boddy, L. (2007). *Fungal Decomposition of Wood*. Springer.

18. Bonito, G., et al. (2010). "Fungal community composition in the rhizosphere of different plant species." *Fungal Diversity*, 43(1), 79-90.

19. Brown, M. E., & Kottke, I. (2009). "Fungal diversity in forest soils." *Journal of Ecology*, 97(2), 255-266.

20. Brundrett, M. C. (2004). "Diversity and classification of mycorrhizal associations." *Biological Reviews*, 79(4), 633-657.

21. Brundrett, M. C. (2006). "Mycorrhizal associations in natural ecosystems." *Plant and Soil*, 286(1-2),

22. 191-208.

23. Carris, L. M., et al. (2007). "Fungal pathogens of plants and their control." *Annual Review of Phytopathology*, 45, 241-268.

24. Casadevall, A., & Pirofski, L. A. (2006). "The fungal cell wall and its implications for pathogenesis." *Clinical Microbiology Reviews*, 19(4), 440-459.

25. Chakraborty, S., & Sinha, R. (2015). "Fungal Flora of Indian Agricultural Soils." *Soil Biology & Biochemistry*, 90, 52-59.

26. Chapela, I. H., et al. (2009). "Soil fungal diversity and functions in ecosystems." *Ecological Applications*, 19(4), 1035-1046.

27. Chattopadhyay, S., & Dey, P. (2012). "Role of Soil Fungi in Nutrient Cycling in Indian Ecosystems." *Ecology and Environmental Sciences*, 38(4), 235-244.

28. Cline, L. (2005). *Fungi in the Soil*. Soil Science Society of America.

29. Colman, S. M., et al. (2011). "Fungal contributions to soil structure and fertility." *Plant Soil*, 339(1), 267-283.

30. Compant, S., et al. (2005). "Fungal endophytes for plant growth promotion." *In: Soil Microbiology, Ecology, and Biochemistry*. CRC Press.

31. Courty, P. E., et al. (2010). "Impact of mycorrhizal networks on soil health." *Soil Biology and Biochemistry*, 42(10), 1750-1760.

32. Covington, W. W., & Moore, M. M. (2006). "Forest fungal communities and ecosystem functions." *Forest Ecology and Management*, 232(1-3), 1-14.

33. Das, P., & Sharma, B. (2016). "Mycorrhizal Fungi and Crop Productivity in India." *Agricultural Research*, 5(2), 119-130.

34. Davidson, E. A., & Janssens, I. A. (2006). "Temperature sensitivity of soil carbon decomposition and feedbacks to climate change." *Nature*, 440(7081), 165-173.

35. Dawson, J. C., et al. (2005). "Soil fungal populations and climate change." *Global Change Biology*, 11(10), 1922-1931.

36. Day, S., & Ainsworth, G. C. (2006). "Fungal interactions and soil health." *Mycological Research*, 110(4), 377-388.

37. Deacon, J. W. (2006). *Fungal Biology*. Blackwell Publishing.

38. Dean, J. F. D., et al. (2009). "Fungal contributions to soil organic matter." *In: Fungal Diversity*. Springer.

39. Deshmukh, S. K., & Pandey, S. (2017). "Fungal Diversity in Indian Desert Soils." *Journal of Arid Environments*, 139, 47-56.

40. Diez, J. R., et al. (2008). "Fungal biodiversity in soils." *In: Soil Microbiology, Ecology, and Biochemistry*. CRC Press.

41. Dighton, J., & Mason, P. A. (2008). "The role of fungi in soil structure." *In: Soil Microbiology, Ecology, and Biochemistry*. CRC Press.

42. Dix, N. J., & Webster, J. (1995). *Fungal Ecology*. CRC Press.

43. Dubey, S., & Sinha, N. (2018). "Impact of Climate Change on Soil Fungal Communities in India." *Environmental Monitoring and Assessment*, 190(6), 375.

44. Edgerton, V. (2009). *Soil Fungi and Soil Health*. Springer.

45. Edwards, I., & Jones, D. T. (2010). "Fungal biocontrol agents and soil health." *Journal of Applied Microbiology*, 109(2), 368-379.

46. Ekelund, F., et al. (2005). "Soil fungi and nutrient cycling." *Plant Soil*, 273(1-2), 91-101.

47. El-Sharif, H. M., & Doughty, D. M. (2007). "Mycorrhizal fungi and soil nutrient dynamics." *In: Soil Microbiology, Ecology, and Biochemistry*. CRC Press.

48. Emmet, A. L. (2002). "Role of fungi in soil health and structure." *Agricultural Research*, 50(1), 34-40.

49. Farina, L., & Aguilar, R. (2006). "Soil fungi and land use." *Agricultural Ecosystems & Environment*, 116(1-2), 45-58.

50. Fisher, P. J., & Hamer, C. J. (2010). "Soil fungal ecology and diversity." *In: Fungal Ecology*. Elsevier.

51. Gadd, G. M. (2001). "Fungal involvement in biogeochemical cycles." *In: The Fungal Community*. CRC Press.

52. Gadd, G. M. (2006). "Fungal contributions to biogeochemical cycles." *Advances in Applied Microbiology*, 61, 29-68.

53. Ghosh, S., & Kumar, A. (2011). "Soil Fungal Populations in Indian Wetlands." *Wetlands Ecology and Management*, 19(2), 115-124.

54. Giller, K. E., et al. (2009). "Soil fertility and fungi." *In: Soil Microbiology, Ecology, and Biochemistry*. CRC Press.

55. Grainger, S. M., et al. (2007). "Fungal endophytes in soil ecosystems." *Mycological Research*, 111(4), 437-450.

56. Gupta, V., & Agarwal, N. (2014). "Soil Fungi as Biocontrol Agents in Indian Agriculture." *Biocontrol Science and Technology*, 24(8), 911-923.

57. Hall, I. R. (2006). *Mycology: A Comprehensive Study*. Springer.

58. Harris, J. A., & Berg, P. (2006). "Fungal community interactions and functions." *Soil Biology and Biochemistry*, 38(4), 840-855.

59. Hibbett, D. S., et al. (2007). "Phylogenetic classification of the Ascomycota." *Mycological Research*, 111(5), 512-520.

60. Hill, D., & Williams, R. (2007). "Soil fungal diversity and ecosystem services." *Fungal Diversity*, 27(1), 79-88.

61. Ingham, E. R., et al. (2005). "Soil fungi and carbon cycling." *In: Soil Microbiology, Ecology, and Biochemistry*. CRC Press.

62. Ingleby, K., et al. (2004). "Functional roles of soil fungi in nutrient cycling." *Journal of Applied Microbiology*, 97(5), 1389-1398.

63. Jackson, R. B., et al. (2006). "Fungal communities in soil and their functions." *Nature Reviews Microbiology*, 4(6), 535-543.

64. Jain, P., & Bhardwaj, N. (2015). "Fungal Adaptations to Indian Forest Soils." *Forest Ecology and Management*, 345, 54-64.

65. Johnson, D., et al. (2006). "Soil fungal diversity and nutrient availability." *Soil Biology and Biochemistry*, 38(6), 1437-1446.

66. Jones, M. D., et al. (2008). "Mycorrhizal fungi in agriculture." *Annual Review of Phytopathology*, 46(1), 223-245.

67. Joshi, A., & Kumar, V. (2016). "Mycorrhizal Fungi and Soil Health in Indian Agroecosystems." *Journal of Soil Science and Plant Nutrition*, 16(2), 321-331.

68. Joshi, M., & Kumar, N. (2017). "Fungal Diversity in Indian Tropical Soils." *Tropical Ecology*, 58(3), 391-400.

69. Karlsson, M., et al. (2013). "The role of fungi in soil organic matter decomposition." *Soil Biology and Biochemistry*, 57, 22-30.

70. Kasuya, M. C. M., & Ferreira, J. A. (2005). "Fungal population dynamics in soils." *Mycological Research*, 109(1), 54-63.

71. Khan, M., & Singh, J. (2018). "Fungal Diversity and Soil Fertility in Indian Agricultural Lands." *Agronomy Journal*, 110(4), 1650-1662.

72. Kjøller, R., & Rosendahl, S. (2007). "Soil fungal communities and ecosystem functions." *Journal of Applied Microbiology*, 103(1), 14-21.

73. Kottke, I., & Hrynkiewicz, K. (2014). "Mycorrhizal networks and soil health." *Fungal Diversity*, 66(1), 103-113.

74. Kumar, S., & Mehta, R. (2013). "Role of Soil Fungi in Organic Matter Decomposition in India." *Journal of Applied Microbiology*, 114(5), 1284-1294.

75. Langerhans, P. B., et al. (2008). "Fungal ecology and nutrient cycling." *In: The Mycota*. Springer.

76. Leake, J. R., et al. (2008). "Arbuscular mycorrhizas and ecosystem functioning." *In: Mycorrhizal Symbiosis*. Academic Press.

77. Lemke, P. A., et al. (2006). "Fungal bioremediation of contaminated soils." *In: Soil Microbiology, Ecology, and Biochemistry*. CRC Press.

78. Lücking, R., et al. (2017). "Fungal taxonomy and ecology." *Fungal Diversity*, 82(1), 1-31.

79. Lücking, R., et al. (2018). "Recent advances in fungal taxonomy." *Mycosphere*, 9(1), 123-144.

80. Martin, J. M., & Moore, R. (2005). "Fungal structure and function." *Mycological Research*, 109(7), 813-823.

81. Martin, K., et al. (2006). "Fungi in soil ecosystems." *In: Soil Microbiology, Ecology, and Biochemistry*. CRC Press.

82. Mehta, S., & Verma, S. (2018). "Mycorrhizal Associations and Soil Quality in Indian Cropping Systems." *Journal of Agricultural Science and Technology*, 20(2), 153-165.

83. Meyer, J. R., & Bailey, L. (2002). "Fungal diversity in agricultural soils." *Journal of Soil Science*, 45(2), 128-140.

84. Mishra, A., & Singh, R. (2012). "Fungal Mycorrhizal Associations in Indian Dryland Soils." *Dryland Agriculture*, 24(3), 145-153.

85. Mohan, M., & Kumar, R. (2017). "Fungal Decomposition of Organic Matter in Indian Soil Ecosystems." *Soil Science Society of America Journal*, 81(6), 1527-1536.

86. Mueller, G. M., & Schmit, J. P. (2007). "Global fungal diversity." *In: Fungal Diversity Research*. Elsevier.

87. Mueller, G. M., & Schmit, J. P. (2008). "Global fungal diversity and the impact of climate change." *Fungal Diversity*, 33, 93-105.

88. Munkvold, G. P., & Marasas, W. F. (2007). "Fungi in soil health." *Plant Pathology Journal*, 56(3), 118-123.

89. Nair, S., & Kumar, S. (2015). "Fungal Responses to Soil Nutrient Variations in Indian Agroecosystems." *Soil Biology and Biochemistry*, 87, 158-165.

90. Nair, V., & Sharma, H. (2014). "Mycorrhizal Fungi and Soil Health in Indian Horticultural Systems." *Horticultural Science*, 49(4), 455-463.

91. Nichols, D. S., et al. (2005). "Fungal metabolic diversity and ecosystem processes." *In: Fungal Diversity Research*. Elsevier.

92. O'Brien, H. E., et al. (2005). "Fungal phylogeny and ecology." *Fungal Diversity*, 20, 1-16.

93. Peay, K. G., et al. (2010). "Soil fungal communities and ecosystem functions." *Annual Review of Ecology, Evolution, and Systematics*, 41, 377-399.

94. Peterson, R. A. (2015). *Soil Fungi and Agricultural Practices*. Springer.

95. Read, D. J., & Perez-Moreno, J. (2003). "Mycorrhizal fungi and ecosystem processes." *In: Fungal Diversity*. Elsevier.

96. Reddy, M., & Reddy, P. (2016). "Fungal Diversity and Soil Fertility in Indian Rice Fields." *Field Crops Research*, 186, 17-26.

97. Richards, T. A., & Bass, D. (2009). "Evolution of fungal pathogens." *Fungal Diversity*, 36(1), 1-13.

98. Roberson, E. B., & Firestone, M. K. (1992). "Soil fungi and soil structure." *Soil Science Society of America Journal*, 56(4), 1382-1390.

99. Rodrigues, J. L., et al. (2015). "Fungal community composition in soil." *Soil Biology and Biochemistry*, 81, 65-76.

100. Sahu, S., & Singh, S. (2013). "Mycorrhizal Fungi in Indian Desert Ecosystems." *Journal of Arid Environments*, 89, 107-116.

101. Sanchez, J. A., et al. (2010). "Mycorrhizal interactions with plant roots." *Journal of Applied Microbiology*, 108(5), 1151-1164.

102. Sangwan, N., & Saini, S. (2017). "Fungal Interactions with Plants in Indian Agroecosystems." *Plant and Soil*, 411(1), 165-179.

103. Schadt, C. W., et al. (2003). "Soil fungal diversity and functions." *Nature Reviews Microbiology*, 1(3), 118-127.

104. Schwartz, D. E., & Vance, C. P. (2002). "Mycorrhizal fungi and phosphorus uptake." *Plant Physiology*, 130(1), 38-45.

105. Siker, A. M., et al. (2006). "Fungal diversity in soil ecosystems." *Journal of Fungal Biology*, 112(3), 181-195.

106. Singh, H., & Kumar, A. (2012). "Ecological Significance of Soil Fungi in Indian Forest Soils." *Forest Ecology and Management*, 263(1), 52-62.

107. Singh, M., & Yadav, A. (2014). "Mycorrhizal Fungi and Soil Health in Indian Agroforestry Systems." *Agroforestry Systems*, 88(1), 133-143.

108. Singh, R., & Jain, P. (2016). "Impact of Soil Fungi on Soil Structure and Health in Indian Agricultural Lands." *Soil & Tillage Research*, 158, 92-100.

109. Sinha, S., & Chaudhary, S. (2016). "Soil Fungi and Their Role in Soil Health Management in India." *Journal of Soil and Water Conservation*, 71(2), 132-140.

110. Smith, S. E., & Read, D. J. (2008). *Mycorrhizal Symbiosis*. Academic Press.

111. Soni, S., & Kumar, P. (2015). "Soil Fungal Dynamics in Indian Coastal Ecosystems." *Marine Ecology Progress Series*, 529, 125-136.

112. Srivastava, R., & Sharma, R. (2014). "Fungal Contributions to Soil Organic Matter Decomposition in Indian Forests." *Forest Ecology and Management*, 323, 123-133.

113. Stajich, J. E., & Hibbett, D. S. (2005). "Fungal genomics and comparative genomics." *Fungal Diversity*, 18(1), 11-23.

114. Stouffer, J. R., et al. (2009). "The role of fungi in soil health and nutrient cycling." *Agricultural and Forest Meteorology*, 149(4), 450-459.

115. Subedi, K., & Patel, A. (2015). "Role of Soil Fungi in Carbon Sequestration in Indian Forests." *Global Change Biology*, 21(9), 2971-2980.

116. Subramanian, P., & Rao, M. (2017). "Role of Soil Fungi in Bioremediation of Indian Contaminated Sites." *Environmental Science & Pollution Research*, 24(7), 6275-6287.

117. Thorn, R. G., et al. (2006). "Taxonomy and ecology of soil fungi." *Mycological Research*, 110(4), 439-451.

118. Tiwari, A., & Singh, B. (2013). "Fungal Biodiversity and Soil Health in Indian Agroecosystems." *Soil & Tillage Research*, 132, 80-91.

119. Tiwari, S., & Gupta, S. (2017). "Fungal Diversity and Soil Fertility Management in Indian Agroecosystems." *Ecological Indicators*, 74, 290-300.

120. Torsvik, V., et al. (2001). "Soil microbial diversity and ecosystem functions." *Soil Biology and Biochemistry*, 33(11), 2131-2140.

121. Vassilev, N., et al. (2006). "Phosphorus solubilizing fungi in soil." *Mycorrhiza*, 16(3), 153-164.

122. Verma, N., & Mehta, A. (2016). "Fungal Adaptations to Soil Nutrient Variability in Indian Agroecosystems." *Agricultural Ecosystems & Environment*, 223, 104-113.

123. Verma, R., & Jain, M. (2014). "Soil Fungal Communities and Soil Health in Indian Grasslands." *Grass and Forage Science*, 69(4), 659-670.

124. Voigt, K., et al. (2009). "Soil fungal communities and climate change." *Soil Biology and Biochemistry*, 41(6), 1214-1226.

125. Wall, D. H., et al. (2008). "Soil biodiversity and ecosystem services." *Nature*, 448(7153), 811-821.

126. Wang, B., & Qiu, Y. L. (2006). "Phylogeny and evolution of mycorrhizal symbiosis." *Mycorrhiza*, 16(3), 265-280.

127. Wang, M. Y., & Wu, G. X. (2009). "Fungal diversity and functions in soil." *Journal of Soil Science and Plant Nutrition*, 9(1), 105-122.

128. Weber, N. S., et al. (2007). "Soil fungi and their role in nutrient cycling." *Environmental Microbiology*, 9(6), 1423-1431.

129. Wiemken, A., et al. (2007). "Arbuscular mycorrhizal fungi and soil health." *Mycorrhiza*, 17(3), 257-269.

130. Wilke, B. M., et al. (2006). "Soil fungi and microbial diversity." *Microbial Ecology*, 51(1), 148-159.

131. Williams, K. J., et al. (2010). "Mycorrhizal fungi in agriculture." *Advances in Agronomy*, 107, 45-78.

132. Yadav, K., & Kapoor, R. (2014). "Fungal Diversity in Indian Urban Soil Ecosystems." *Urban Ecosystems*, 17(2), 381-392.

133. Yadav, N., & Sharma, S. (2015). "Mycorrhizal Fungi and Soil Nutrient Dynamics in Indian Croplands." *Journal of Soil Science and Plant Nutrition*, 15(4), 1101-1110.

134. Yadav, S., & Kumar, S. (2018). **"Fungal Responses to Soil Management Practices in Indian Agroecosystems."

135.Yang, Z., & Xu, J. (2008). "Fungal diversity and ecosystem functions." *Ecological Applications*, 18(6), 1634-1646.

136.Zhang, Y., et al. (2007). "Soil fungal community dynamics." *Soil Biology and Biochemistry*, 39(5), 1234-1243.

137.Zhao, Q., et al. (2005). "Fungal bioremediation of environmental pollutants." *Environmental Science and Technology*, 39(12), 4244-4250.

138.Zhao, X., & Patel, J. (2016). "Soil Fungal Communities in Indian High-altitude Ecosystems." *Journal of Mountain Science*, 13(8), 1382-1391.

139.Zheng, Z., et al. (2011). "Metagenomics and soil fungal communities." *Soil Biology and Biochemistry*, 43(5), 1134-1141.

140.Zilli, J., et al. (2009). "Fungal diversity and environmental impact." *Fungal Diversity*, 37(1), 107-118.

Appendix A: Additional Resources

This appendix includes a curated list of additional resources for readers seeking to expand their knowledge of soil fungi. It contains books, research journals, online databases, and organizations dedicated to mycology, soil science, and ecology.

1. **Books**

2. *Fungal Biology* by J.W. Deacon – A comprehensive textbook on the biology of fungi, covering everything from cellular processes to fungal ecology.

3. *Mycorrhizal Symbiosis* by Sally E. Smith and David J. Read – A definitive text on the biology of mycorrhizal fungi and their symbiotic relationships with plants.

4. *Fungi in Biogeochemical Cycles* edited by Geoffrey M. Gadd – Focuses on the role of fungi in biogeochemical cycles and environmental sustainability.

5. **Research Journals**
   - *Fungal Ecology* – A journal focusing on the ecological roles of fungi in various environments.
   - *Soil Biology & Biochemistry* – Publishes research on the biological and biochemical processes in soils, including those mediated by fungi.
   - *Mycologia* – The official journal of the Mycological Society of America, covering all aspects of fungi.

6. **Online Databases**
   - **UNITE** (Unified system for the DNA-based fungal species identification) – A key resource for identifying fungal species based on DNA sequences.
   - **FungiDB** – A genomic and functional database that provides information on fungal species and their genes.
   - **MycoBank** – An online database that provides taxonomy and nomenclature information for fungi.

7. **Organizations**
   - **The Mycological Society of America** – A society dedicated to promoting the study of fungi, with resources for mycologists and educators.
   - **International Mycorrhiza Society** – Promotes research and education on mycorrhizae and their application in agriculture and forestry.

Appendix B: Data Tables

This section provides detailed data sets related to fungal diversity, soil composition, and ecological functions. The data can be used to support research, comparative analysis, and teaching.

1. **Table 1: Soil Fungal Biomass Across Different Ecosystems**

Ecosystem	Fungal Biomass (mg/g soil)	Dominant Fungal Group
Temperate Forests	8.7	Ectomycorrhizae
Tropical Rainforests	6.5	Saprotrophic Fungi
Grasslands	4.2	Arbuscular Mycorrhizae
Agricultural Soils	3.1	Saprotrophic Fungi
Boreal Forests	9.3	Ectomycorrhizae

2. **Table 2: Nutrient Release by Fungi During Decomposition**

Nutrient	Fungal Enzyme Responsible	Average Release Rate (μg/g/day)
Nitrogen (N)	Protease	3.5
Phosphorus (P)	Phosphatase	2.8
Potassium (K)	Potassium-solubilizing Fungi	1.2
Carbon (C)	Cellulase, Ligninase	5.7

3. **Table 3: Common Mycorrhizal Fungi and Their Host Plants**

Fungal Species	Host Plant Family	Fungal Type (Endo/Ecto)
Glomus intraradices	Grasses, Legumes	Endomycorrhizae
Boletus edulis	Pine, Oak	Ectomycorrhizae
Rhizophagus irregularis	Corn, Wheat	Endomycorrhizae
Pisolithus tinctorius	Eucalyptus, Pine	Ectomycorrhizae

Appendix C: Extended Discussion on Mycorrhizal Networks

In this section, we delve deeper into the concept of the "Wood Wide Web," the vast underground network of mycorrhizal fungi that connects plants and facilitates communication, nutrient sharing, and defense mechanisms.

1. **Nutrient Transfer in Mycorrhizal Networks**

   Mycorrhizal networks enable plants to share nutrients, especially in ecosystems with uneven nutrient distribution. Fungi play a central role in transferring carbon, nitrogen, and phosphorus between plants, ensuring that weaker or younger plants can access resources. Research shows that through these networks, plants can also exchange water during drought periods, highlighting the collaborative nature of plant-fungal relationships.

2. **Chemical Signaling and Defense**

   Plants connected by mycorrhizal fungi can communicate about environmental threats, such as herbivory or pathogen attack. Through the network, a plant under attack can release volatile organic compounds (VOCs) that signal neighboring plants to activate their defensive mechanisms, such as producing toxins or strengthening cell walls to ward off pests.

3. **Fungi as Ecological Mediators**

   The mycorrhizal network also mediates competition between plants by redistributing nutrients, which can either alleviate or exacerbate competitive dynamics. Dominant plants sometimes "pay" the fungi in carbon to preferentially deliver nutrients to themselves, while other times the network balances the distribution in favor of overall ecosystem health.

Appendix D: Extended Discussion on Fungal Roles in Carbon Sequestration

This section provides an in-depth discussion on how fungi contribute to carbon cycling and sequestration in soils, a critical component in mitigating climate change.

1. **Fungi in the Carbon Cycle**

   Fungi are essential players in the global carbon cycle, breaking down organic matter such as dead plant material and facilitating its conversion into humus. During decomposition, saprotrophic fungi release carbon dioxide (CO_2) back into the atmosphere, but they also contribute to long-term carbon storage by converting organic carbon into stable forms like humic substances.

2. **Mycorrhizal Fungi and Carbon Sequestration**

   Mycorrhizal fungi, particularly ectomycorrhizal species, enhance carbon sequestration by slowing down decomposition. They form strong associations with plant roots, reducing the availability of plant litter for saprotrophic fungi, which in turn limits the release of carbon as CO_2. This interaction locks carbon in the soil, contributing to carbon storage in forest ecosystems.

3. **Potential for Climate Change Mitigation**

   By promoting fungal growth and health, land managers can enhance the soil's capacity to sequester carbon, providing a natural solution to climate change. For example, promoting mycorrhizal associations in reforestation projects not only helps trees grow faster but also increases the soil's carbon-holding capacity, contributing to long-term climate mitigation efforts.

Appendix E: Extended Discussion on Fungal Pathogen Control in Agriculture

This appendix explores how fungi are used as biological control agents to manage soil-borne plant pathogens in sustainable agriculture.

1. **Fungi as Biocontrol Agents**

   Fungi like *Trichoderma, Penicillium,* and *Beauveria* have been widely used as biocontrol agents to suppress harmful pathogens in agricultural systems. These fungi outcompete pathogens for resources, produce enzymes that degrade pathogenic cells, or induce plant defenses, reducing the need for chemical fungicides.

2. **Mechanisms of Pathogen Suppression**

   Biocontrol fungi employ multiple mechanisms to suppress plant diseases. *Trichoderma,* for example, produces chitinases that break down the cell walls of fungal pathogens. It also forms a protective barrier around plant roots, preventing pathogens from accessing the plant tissues.

3. **Challenges and Future Prospects**

   While biocontrol fungi hold great promise, their effectiveness can be influenced by soil conditions, temperature, and the presence of other microorganisms. Research is ongoing to develop more resilient strains and improve their application methods for widespread agricultural use.

Endnote

As we reach the conclusion of *The Hidden World Beneath: An Exploration of Soil Fungi*, it becomes evident that the world beneath our feet is far more complex, interconnected, and crucial to life on Earth than we often realize. Soil fungi, these often unseen and misunderstood organisms, form the foundation of ecosystems and influence everything from nutrient cycling to plant health, soil structure, and even climate stability. Their contributions stretch across forests, grasslands, agricultural systems, and urban environments, shaping the very fabric of terrestrial life.

Throughout this journey, we have delved into the fascinating symbiotic relationships fungi establish with plants, especially through the mycorrhizal networks. These networks are not merely conduits for the exchange of nutrients and water; they are also sophisticated communication systems that allow plants to respond to environmental stressors, pathogens, and even the needs of neighboring plants. The cooperative and mutually beneficial nature of these fungal networks reveals an intricate balance in nature—one where competition is complemented by cooperation, and survival is often a collective endeavor.

Our exploration has also highlighted the role of fungi in maintaining soil health and biodiversity. As primary decomposers, fungi are essential in breaking down organic matter, releasing critical nutrients back into the soil, and facilitating the formation of humus, which contributes to soil fertility and structure. The richness of fungal biodiversity in soils ensures resilience against environmental changes, promotes plant diversity, and mitigates the impact of soil-borne diseases. It is through their constant activity that ecosystems remain productive and sustainable, often in the face of challenges such as nutrient scarcity and changing climate conditions.

Moreover, the role of fungi in plant immunity and pathogen resistance underlines their importance in agriculture and forestry. Fungi not only prime plants to defend themselves against pathogens but also act as biological control agents, suppressing the spread of harmful organisms. This aspect of fungal biology opens up new possibilities for sustainable agricultural practices that reduce reliance on chemical inputs, enhancing the health of crops while preserving the natural integrity of ecosystems.

As we step away from this subterranean world, we are reminded of the vital ecological significance of soil fungi in promoting resilience in both natural and managed ecosystems. From enhancing carbon sequestration to fostering drought resistance and supporting plant community dynamics, soil fungi continue to reveal themselves as key players in the global environmental system.

Yet, much remains to be discovered. The science of soil fungi is still evolving, with ongoing research uncovering new species, interactions, and functions that were previously unknown. Our current

171

understanding is but a glimpse into a vast, hidden world that holds immense potential for addressing the environmental challenges of the future. As we learn more about fungi, we also deepen our appreciation for the delicate balance they help maintain in the web of life.

In closing, *The Hidden World Beneath* invites us to look below the surface and appreciate the profound influence that soil fungi exert on our world. Their work may be hidden from view, but their impact is felt everywhere—from the health of plants in our gardens to the sustainability of forests, the productivity of farmlands, and the future of our planet's ecosystems. It is our hope that this book sparks curiosity and a greater recognition of the importance of soil fungi in shaping the living world, encouraging further exploration, conservation, and stewardship of the unseen life that lies beneath.

Let us continue to dig deeper, both in knowledge and in practice, to protect and sustain the remarkable fungal networks that sustain life on Earth.

Dr. Ravindra Goswami
Dr. Vishwajeet Singh
Dr. Anita Chauhan
Dr. Rashmi Vamil
Prof. Seema Bhadauria